Low-Carbon Birding

Low-Carbon Birding

Edited by Javier Caletrío

Pelagic Publishing

First published in 2022 by
Pelagic Publishing
20–22 Wenlock Road
London N1 7GU, UK

www.pelagicpublishing.com

Low-Carbon Birding

British Library Cataloguing in Publication Data
A catalogue record for this book is available from the British Library

ISBN 978-1-78427-344-6 Hbk
ISBN 978-1-78427-345-3 ePub
ISBN 978-1-78427-346-0 PDF

https://doi.org/10.53061/EPEA5466

Cover image © Gary Redford

All interior illustrations © Alan Harris

Typeset in Adobe Garamond Pro by Deanta Global Publishing Services, Chennai, India

Printed in England by TJ Books

Nothing is enough for the man to whom enough is too little.

Epicurus

Maximize meaning, minimize carbon, is one measure of a life well lived.

Kimberly Nicholas

Contents

Contributors

Nick Acheson lives in a flint cottage beside the River Wensum in north Norfolk. He grew up nearby, beside the River Stiffkey. In between moving the three miles from one north Norfolk river to another, he spent ten years in South America (largely by the rivers Piraí, Iténez and Mamoré in Bolivia), four in India (along the Brahmaputra, among many others) and worked with wildlife on every continent. Having reflected deeply on climate, biodiversity and ecotourism, he has given up flying. He now cycles around north Norfolk, and walks along the Wensum. He writes, speaks and teaches about wildlife and conservation, often on behalf of Norfolk Wildlife Trust and Felbeck Trust.

Tim Allwood is a low-carbon birder and teacher. After a period working and birding overseas, he moved to Norwich in 2001 before finally making it to the Norfolk coast in 2007 and adopting his beloved patch of TG42. Besides obsessively searching for birds in his local area, he grows vegetables, cycles locally, kicks the occasional football and obsesses over music. He is now most often found walking his greyhound, riding his bike or working his allotment. He does his best to live and promote a low-carbon lifestyle.

Mark Bannister is married with a son and daughter. He worked as an aerodynamicist specialising in wing design but never even got close to designing anything to match the aerodynamics of a swift. After a career in which much time was spent gazing out of the window, he is now retired and spends as much time as possible outside in his garden-come-nature-reserve with his wife, Mandy, or out on his bike birding the Humber bank and the tracks and minor roads of north Lincolnshire. Far too late in the day—but better late than never—he is decarbonising his lifestyle and unexpectedly thoroughly enjoying it.

Amanda Bourne is a conservation planning officer with the Northern Agricultural Catchments Council in Western Australia. Originally from South Africa, she completed her PhD at the Fitzpatrick Institute of African Ornithology in Cape Town in 2020. Her doctoral research focused on the impacts of climate change on the behaviour, breeding success and physiology of southern pied babblers. She is particularly keen on avian ecology and understanding and responding to the impacts of climate change in dryland environments.

Javier Caletrío has a background in economics and sociology and is a researcher in the areas of travel, inequality and ecological transitions. He is currently research advisor to the Mobile Lives Forum, a research institute funded by SNCF (French railways). He has lived in England for 22 years and travels every summer to the east coast of Spain by train via London, Paris and Barcelona to see his family and watch birds in his former patch in l'Albufera de Valencia. He cofounded the Valencian annual bird report and the naturalist group Roncadell.

Jonathan Dean made his first birding trip on 30 December 1991, aged nine, to the Eden Estuary in Fife, and he has been a keen birder ever since. His main birding claim to fame is that in the 1990s he was three times winner of *British Birds* magazine's Young Ornithologist of the Year competition. He now devotes most of his energy to patch birding in and around his home city of Coventry, West Midlands, with occasional forays further afield. A political scientist by profession, he takes a keen interest in the social and political dimensions of birds, birding and nature conservation.

José Ignacio (Nacho) Dies Jambrino is a birder and wetland manager born and raised in Valencia, Spain. He has been captivated by natural history and dedicated to bird census and ringing activities since he was a child. He is currently involved in the management of the Racó de l'Olla reserve, as well as in the monitoring and conservation of bird populations in l'Albufera de Valencia. He was the cofounder of the Valencian annual bird report in the late 1980s and, among other activities, has served as a member and secretary of the Spanish Rarities Committee.

Steve Dudley is an obsessive birder and garden lister. He has recently relocated from the Cambridgeshire fens to the north Orkney isle of Westray as part of his drive to lead a lower-carbon lifestyle, which now includes no flying. He has worked in bird conservation and ornithology for 37 years, with the Royal Society for the Protection of Birds, the British Trust for Ornithology and, for the last 25 years, has run the British Ornithologists' Union. Early retirement in 2022 has allowed him to become a full-time birder. Steve has authored three books: *Rare Birds Day by Day* with Tim Benton, Pete Fraser and John Ryan; *Watching British Dragonflies* with Caroline Dudley and illustrated by Andrew Mackay; and *A Birdwatching Guide to Lesvos*.

Arjun Dutta has been interested in birds and wildlife since he was seven. Now 19 and studying geography at the University of Cambridge, he has been lucky enough to share this interest with a range of people and organisations. He currently represents the Cameron Bespolka Trust as youth ambassador and he sits on the Youth Advisory Panel for the British Trust for Ornithology. In addition to working as a young volunteer for the National Trust in south London, much of his spare time is spent sound recording British birds; his recordings are often shared using a Twitter account called Ichos that he set up with a group of friends, which aims to help more people learn about the power of bird sound.

Roger Emmens has had a lifelong interest in birds and, indeed, nature more generally. Now retired, after a career in information technology, he volunteers with various bird study and conservation projects locally in the Epping area. He has been a ringer since his youth—he ringed his first bird, a starling, in Skegness in 1965. Nowadays, he is an active member of the Rye Meads Ringing Group.

Jonnie Fisk has lived and worked at Spurn Bird Observatory on the Yorkshire coast for the last five years, after first visiting the area in his teens. He enjoys the company of the Humber estuary's brent geese and grey plovers and lives vicariously through the journeys of these and other avian migrants. He is currently busy waiting to see his first nuthatch at Spurn, arranging and rearranging his notebooks, and trying to shrink his horizons to his pushbike-able local area while contemplating his place in the world.

Steve Gale has been birding across north Surrey and south London for almost 50 years, with much of that time spent at Beddington Sewage Farm, Holmethorpe Sand Pits, Canons Farm and, especially, the scarp and dip slopes of the North Downs; he can sometimes be found in nearby Sussex and Kent. A fascination with migration has led him to become addicted to skywatching. Patch-working has always been a part of his birding ethos, and he likes nothing better than to search out nearby areas with little ornithological coverage. This has led to some notable discoveries, particularly that of the UK's largest ever recorded gathering of hawfinches, in 2018. He is also a keen recorder of insects and plants.

Simon Gillings is a keen birder, artist, low-carbon lister and addicted to monitoring nocturnal bird migration ('nocmig'). Originally from Lincolnshire, he now lives in Cambridge where he misses the sea. Since the mid-1990s he has worked at the British Trust for Ornithology on various monitoring and research projects. In his current role as principal data scientist he supervises research and development for the Breeding Bird Survey, BirdTrack and the BTO Acoustic Pipeline.

Gavin Haig cut his birding teeth during the early 1980s, when he could frequently be found on the causeway at Staines Reservoirs in West London, or chasing rarities around the UK as a keen twitcher. He is much older now, and slower, and is happiest when birding the coast near his home in West Dorset. He particularly enjoys spring migrants, tricky gulls, and the very occasional thrill of a nice find. He also enjoys writing—mostly about birding—and publishes a regular blog.

Finley Hutchinson is an 18-year-old entomologist, birder and general wildlife enthusiast from Berkshire. He has been interested in nature from a young age, taking up birding several years ago and entomology at the start of March 2020. Since then he has spent all his free time surveying and identifying, and has written academic papers on his findings. Finley hopes to study conservation and ecology at university from September 2022.

Dave Langlois has been a cycling birder for over 60 years now, first around the unprepossessing scenery of Hayes, Middlesex, where he was born, and then in the more congenial countryside of Kent. In 1990 he moved to Spain where he continued pedalling, initially around Madrid and latterly in Extremadura and Asturias. He is author of three nature-based novels in Spanish and another book on birdsong, his special interest and delight. Among many other jobs he has been assistant warden at Stodmarsh Nature Reserve.

Kieran Lawrence is a 25-year-old birder, currently studying for a doctorate in biological sciences at Durham University. A passion for birds has taken up most of his life, starting with watching red kites while on a camping trip to Dumfries with his father at the age of eight, and steadily growing into a full-blown obsession. He is a bird ringer, spending much of each year at Spurn Bird Observatory, a migration hotspot on the Yorkshire coast.

Alexander Lees is a senior lecturer in conservation biology at Manchester Metropolitan University and associate of the Cornell Lab of Ornithology. He spent his adolescence trying to find rare birds at a sewage farm in Lincolnshire, after which he moved on to spend 18 years undertaking research in various far-flung corners of the earth—including five years living in the Amazon. He is now back in the UK trying to find rare birds in one of the least likely locations in the UK—a valley in the Peak District.

Sorrel Lyall is a birder, artist and trainee ringer from Nottingham, currently living in Edinburgh where she studies ecology. Her obsession with birding began, aged nine, through exploring local nature with her birding grandparents. With a keen interest in youth engagement, Sorrel is President of the Edinburgh University Ornithological Society and was a member of the British Trust for Ornithology Youth Advisory Panel. Sorrel has recently spoken out about issues of diversity and inclusivity in the conservation sector and campaigns to increase engagement of minority groups in birding.

Siân Mercer is an 18-year-old birder and environmentalist from north Shropshire, interested in everything from plastic pollution to peat bogs. She is involved in a number of surveys and has created small-scale projects in her local area to raise awareness of plastic pollution, through school and with friends. She is passionate about her work as a British Trust for Ornithology youth representative, helping to engage more young people with the wildlife around them and to inspire the next generation.

Lowell Mills-Frater is a first-generation birder from Newcastle upon Tyne, UK. He was inspired to get into birding by an early passion for nature, time with his grandparents, Patrick and Anne Mills, in the Wye Valley, family day visits to the Farne Islands and primary school trips to Thornley Woodlands and Big Waters. Having studied zoology as an undergraduate at Newcastle University and an MRes in ecology at the University of Glasgow, he completed his PhD on potential breeding-ground factors in the decline of the cuckoo at Exeter University in 2019.

Nick Moran was born in Yorkshire and now lives in south-west Norfolk where he is the British Trust for Ornithology's training manager. Nick's work involves designing and running activities to help people improve their bird identification skills and participate in BTO surveys. He draws on ten years' experience of teaching biology in the UK, Uganda, China and the UAE. Birds have been Nick's lifelong passion—some have said 'obsession'—and he now puts most of his energy into local birding, either on his beloved patch (BTO's Nunnery Lakes) or further afield in East Anglia, usually by bike.

Matt Phelps is an avid birder, naturalist, land manager and writer, originally from Hampshire, now based in West Sussex, and often found outdoors. Matt's fascination with the natural world began with childhood explorations of local woodlands and heathlands. He has spent the bulk of his working life nurturing landscapes for the benefit of wildlife, while also carrying out ecological surveys and various other nature-related activities, such as leading guided walks and talks. Among other things, he is currently an eBird reviewer for Sussex and sits on the Sussex Ornithological Society Records Committee.

David Raffle is a young birder, naturalist and conservationist. His time is split between his home in Newcastle and Cornwall, where he is studying conservation biology and ecology at the University of Exeter's Penryn Campus. In his spare time he volunteers for local conservation organisations and is a youth representative for the British Trust for Ornithology. He is passionate about low-carbon birding and watching wildlife locally and in a more sustainable way. He is most often found on his local patch, Gosforth Nature Reserve, looking out for birds and insects.

Amy Robjohns is a lifelong Hampshire resident, keen birder, biological recorder, and assistant recorder for Hampshire Ornithological Society. She graduated from the University of Southampton with a BSc in environmental sciences and was awarded the Dawn Trenchard Prize for her enthusiasm and dedication to her studies. She continues to channel most time and energy into her beloved local patch, Titchfield Haven, or exploring Scottish islands. Amy works at the Hampshire Biodiversity Information Centre as an ecologist focusing on digitising detailed habitat data, data requests (species data, habitat data and protected sites information), and screening planning applications.

Maria Scullion is a birder and all round nature lover from North Yorkshire, now living in Nottinghamshire. Her love of the outdoors started at a young age when her nursery teachers commented on her love of woodlice; she was further influenced by her family and beautiful surroundings as she grew up. She studied animal science at the University of Nottingham, which included a jam-packed year in industry working at the Wildfowl and Wetland Trust's Martin Mere reserve. In her free time she volunteers for the British Trust for Ornithology as a youth representative and also enjoys bird ringing.

Ben Sheldon is Luc Hoffmann Professor of Field Ornithology in the Department of Biology at the University of Oxford. A keen birder and ringer since his childhood, spent in Hampshire and Norfolk in the 1980s, he really cut his birding teeth at UK bird observatories, particularly Portland Bill in Dorset. He embraced ornithology as a research subject after undergraduate studies at Cambridge University, via a PhD at the University of Sheffield and postdoctoral studies in Uppsala (Sweden) and Edinburgh. Since moving to the University of Oxford in 2000 his research has focused on long-term population studies of tits in Wytham Woods. Local birding and back garden moths have enabled him to keep his natural history interests alive while serving as head of the Department of Zoology.

Angela Turner is managing editor of the journal *Animal Behaviour* and is on the Notes panel of the journal *British Birds*. Her fascination with hirundines stems from her doctoral work on the foraging behaviour of these birds. As well as many scientific papers and popular articles, she has written several books including *A Handbook to the Swallows and Martins of the World* (co-authored with Chris Rose) and *The Barn Swallow*. Originally from London, she now lives in Nottingham, UK.

Steven Ward was born and raised in the northern Yorkshire Dales, and is still there now, 40 years later. The birds must be good! Living in a rural area, he was drawn to nature from a young age, spending much time wandering the fields down by the river. His interest really took off in his twenties when he began monitoring peregrines and ravens for the local raptor group. Now, with two young children and a little less time on his hands, he focuses on patch and low-carbon birding in his local area.

Foreword

When I finished as CEO of the Royal Society for the Protection of Birds, lots of people asked what trips I was now going to go on. It seemed to them to be the obvious thing that I would do, now that I had the time and opportunity. As a teenager with an ambition to work for the British Antarctic Survey, my imagination was gripped in 1985 by Martin Saunders' and Mike Salisbury's remarkable footage of belugas, narwhals and little auks teeming at the edge of the food-rich ice shelf as it receded in the Arctic spring, from the BBC's *Kingdom of the Ice Bear*. My heart was set on the ice edge in Lancaster Sound, by Prince Leopold Island, as *the* place in the world where I most wanted to go.

It still is. But I have known for some while that I cannot go. It would break my heart. It would be too raw—the signs of global heating are magnified wherever you go in the Arctic—and I would never be at peace with myself. I have been on a journey of adjustment for some while. What the Covid-19 pandemic has taught me is that I am fine with that. Slowing down and staying local has reopened windows of childhood wonder, long closed in the mad rush of my adult life.

In the brilliance of the 'lockdown spring' of 2020, I discovered and shared the intricate world of #pavementplants. For the first time in years, I opened my copy of C. E. Hubbard's *Grasses* (a natural history classic) and my daily walk became a richer experience as I could recognise all the species, previously ignored underfoot. My horizons expanded massively with exciting new wildlife encounters right here at home—thanks to an infra-red trail camera, a state-of-the-art bat detector and a simple, plastic digiscope phone adaptor (all costing much less than birding optics or even a tripod). There is so much more in the natural world here around me that I plan to discover.

Bringing a wide range of essays together in this book, Javier shows the varied and enriching ways in which it is possible to be inspired by wildlife and be part of nature, to find a sense of place, and to care for nature and people. By describing birdwatching journeys through life, the authors help address questions that a growing number of birders and others inspired by nature are asking themselves:

How do I stay connected with the nature that I really want to keep in my life?
What do I need to let go of—in order to not make matters worse?
What can we help to bring back?

These reflect much deeper questions of adaptation that may well become the defining questions of our age for millions who live in societies built on the energy of fossil carbon.

Humanity is part of the biosphere—*the living part of Earth*. Nature is inside us as much as we are inside nature. Human wellbeing in all its forms is built fundamentally on nature's capacity and inter-dependence with the Earth's life support systems.

There is every likelihood that an ice-free Arctic will occur one summer in the next few years. Ecological disruption is underway. Due to our activities, the biosphere is destabilising rapidly. The climate/nature crisis is not a medium-term issue about the Global North's future, it is present now in the lives of many in the Global South. The debate over eco-tourism is in danger of being consigned to history—arguably a human-induced pandemic has just done so—and the institutional inertia to address this issue may soon look like denial.

The scale of the threats to the biosphere is, in reality, so great that it is difficult to grasp even for well-informed experts.

Industrialised cultures have failed to recognise humanity's power as a new, dominant force in the operation of the biosphere and how it shapes Earth's systems more broadly. The evidence is increasingly compelling that future environmental conditions will be far more dangerous than is currently believed by mainstream opinion.

There is going to be a new normal—it is simply a question of what kind of normal. We cannot continue to consume more than the Earth has capacity to regenerate. Over geological time, the Earth will recover from planetary overshoot; the choice is whether human civilisation stays the course. The challenge for all of us now is to understand how to effect deep, systemic change fit for a new reality.

This book is about more than helping us to reimagine our lives in the new normal society we want to live in. It is also about you and me taking power to help make it happen. We cannot change the reality of the planetary emergency here and now, but the outcome will depend on how we all choose to respond.

Vaunted as the pivotal event for the decade, COP26 was yet another demonstration that politicians remain impotent to bring about the transformative changes in society we need. And that should not be a surprise, because the power in society to change shared beliefs, attitudes and behaviours comes from individual citizens and civil society. Governments and businesses do not lead; they follow.

I have made only a small start in the changes I know I have to make in my life, and that will need to be made in the society around me to enable me to continue to adapt. I know I have a long way to go and that I will need to change faster, but I am happier on this journey—and there is so much nature around me and so little time to explore it all.

It seems to me, now, that it would be a pretty good deal just to be able to live in a world where I can go to sleep at night secure in the knowledge that there will always be Arctic sea-ice—*and* dream.

Mike Clarke

Preface

The way in which many birdwatchers in the UK have come to understand and enjoy their hobby can only be fully understood in a context of rampant overconsumption of fossil fuels during the last four decades. But like many other aspects of our societies, birdwatching is already changing in response to the climate crisis.

I first realised that this change was happening in spring 2018. In April that year the magazine *British Birds* published a short article that I had written to highlight the need to move towards a birdwatching culture that acknowledges the reality of the climate crisis, the required scale of mitigation and the need for lifestyle change. After its publication, I began receiving enthusiastic emails from readers happy to see the subject finally being addressed. *British Birds* was also receiving such emails. In May 2018 Roger Riddington told me that, as far as he could remember, no editorial since he had become editor in 2001 had received so many comments from readers, all positive except for two, which were published in the June issue along with my response. It was encouraging to know that there was a latent desire for change.

This article in *British Birds* and a website that I created in 2020 helped to encourage some birdwatchers to speak publicly about their desire for a different approach to birding and their own personal experiences of adopting climate-friendly ways of enjoying birds. These are stories of change that I think deserve to be known, in the hope that they will encourage others to share their own experiences. This book is a compilation of some of those stories.

These are simply personal accounts, not blueprints for low-carbon birding. What exactly birdwatching could look like on a liveable planet is something that we will learn as we muddle through the crisis. It will depend on decisions made by each society about future transport systems; the way we organise where and how we live, work and consume; our views on inequality; and how much we let nature regenerate in and around cities and in the countryside. Importantly, it will depend on who makes these decisions. All this remains uncertain at the moment.

What the chapters in this book do offer, however, are hints of where we could begin as we search for a different birdwatching culture. In many of these accounts that search starts with the ordinary birds and ordinary routines of everyday life itself—routines and birds that bring solace, comfort and meaning to many. What transpires from many of these stories is that every birding day, no matter how ordinary, can be a festive day; that wonder can spring forth from the immediate environment surrounding us; and that the ordinary can also be fulfilling, even magical. Perhaps this is not a bad place to start, considering that tales about ordinary birds are routinely shared by millions of people who care for what is good about their neighbourhoods, villages, towns and cities across the world. These accounts facilitate connections that nurture shared experiences and identities.

The accounts in this book are not about heroic efforts to save the planet. They are simply stories of humble everyday humanity in unprecedented times—ordinary people with doubts and concerns about how to live a decent life and act responsibly in a warming world. I hope the reader will appreciate the authenticity of their voices.

North Yorkshire, 2022

Acknowledgements

Following my arrival in the UK in 1998 my birdwatching was always a rather solitary pastime. It still is—I enjoy solitude. But since 2018 I have also known that every day I have gone out on a walk around the Lune estuary or on a birding day out by train around Morecambe Bay or the Lake District, many others around the country were also walking paths, riding bicycles, jumping on buses and enjoying passing landscapes from the train, small gestures sending a clear message that a different birdwatching culture is possible. Some of them had been doing it and talking about it for more than a decade. Many thanks to those who have shared their experiences with me over the last three years. Special thanks also to all the contributors in this book and those who for various reasons could not be part of it. These are your stories and I hope the book will encourage you to continue talking about the climate crisis. Many thanks to Nick Acheson, Nacho Dies Jambrino, Steve Dudley, Steve Gale, Gavin Haig and Matt Phelps who kindly provided useful comments and suggestions about an earlier version of the introduction, and to Nigel Massen from Pelagic Publishing for his interest in making this book a reality. My wife, Angela, has made my birdwatching and my life much more interesting with her disarming questions about ordinary birds. She has been involved in every stage of editing this book and has been a most perceptive reviewer of all the chapters.

Introduction

Javier Caletrío

There is beauty in the simplicity of birdwatching. A pair of binoculars and a curious mind are all you need to enjoy a pleasant and rewarding pastime. It is not surprising then to find around the end of November birdwatchers posting images of recent sightings on social media as a way to express their disdain for the materialist ethos associated with Black Friday. Interestingly though, these posts reveal a blind spot in the way birdwatchers often conceive of their hobby. While most of these photos are from areas local to the person posting, others are of birds from distant places. In 2020 one such post included photos of birds and mammals taken by a birdwatcher on five continents. Another showed a colourful passerine from Australia. The well-intentioned English birdwatcher posting that photo was probably unaware that her return flight to Sydney had generated the equivalent of 480 suitcases of carbon, each case weighing 20 kilograms.[1] Just imagine coming back home from your wildlife holiday, the smile on your face reflecting pride in what you think is a lifestyle revolving around experiences rather than material possessions, and finding 480 cases filled with carbon in your own back garden. Flying might not entail an accumulation of material stuff in our houses, but we instead chuck carbon through the plane window at a rate of 50 bricks per hour, each brick weighing 2.4 kilograms. And each of our fellow 350 passengers in that flight is doing the same. Pouring that carbon into the local wetland would be unthinkable, and yet we have come to live with the illusion that the atmosphere is not part of nature or that it is just a limitless sink. Birdwatching can be light and simple but it can also be heavily polluting.

People enjoying high-carbon lifestyles that depend on frequent flying and long-distance driving are making a disproportionate contribution to climate change, and the problem is that there is no technological fix that can generalise for 7.9 billion people the material standard of living currently enjoyed by a minority at a high cost to others.[2] The key issue, however, is not that too many birdwatchers routinely engage in long-distance driving and frequent flying, but that wildlife media and influential organisations in the birding world often celebrate these styles of high-carbon birdwatching. And if this kind of behaviour is regarded as normal and acceptable by those who are supposed to know and care for nature, then what message are birdwatchers and conservationists sending to the rest of society? What sense of urgency are we conveying? The challenge is not one of changing individual minds but of cultural change. We need to normalise behaviours and worldviews that are aligned with the realities of the climate emergency. We need to reimagine birdwatching for a liveable planet.

I say reimagine because, beyond the need to reduce fossil fuel demand, we must also reassess how we value birds and places and, ultimately, how we conceive of ourselves and our relation to the world. There is a rich diversity in the experiences we can have here and now, but we must adjust our values and lives to see it that way. Whether it is marvelling at the playful flight of the swallow, patiently listening to redwing calls on a calm autumn night, sharing a cup of tea after fieldwork and discussing unanswered questions about the birds in the patch and beyond, or savouring the first glimpse of a nutcracker from an Alpine train, the chapters in this book show that we can enjoy diverse and meaningful experiences of birds while consuming, or at least making a reasonable effort to consume, no more than an equitable share of nature. Birdwatching can and should be redefined around a commonly held sense of what is enough.[3]

The good news is that key elements of what might constitute a low-carbon birding culture today are already there. Patch birding has a long and respected tradition going back to Gilbert White's studies of Selborne, Horace Alexander's pioneering surveys of breeding birds, Emma Turner's birdwatching on Scolt Head, and David Lack's classic study of the robin. And this continues today in, for example, the recent *Sound Approach guide to the Birds of Poole Harbour*, John Threlfall's portraits of the Solway Firth, and the collective efforts orchestrated by the British Trust for Ornithology and many local groups such as the Rye Meads Ringing Group to monitor trends in bird populations. In addition, many people are beginning to explore the possibilities of low-carbon travel for their birdwatching trips and holidays.

But we need to be more confident in viewing the commitment to enjoying birds more locally and low-carbon travelling not as second best options for those who cannot afford or choose not to own a car, drive long distances or fly frequently. A culture of moderation is something to celebrate. Not only is it a sensible thing to do, it can open our eyes to an abundant world. As Ben Sheldon argues in this book, fulfilling birdwatching experiences do not have to involve long lists of birds. There is, he argues, 'diversity in behaviour, in individual lives, in parasites, songs, learning, social relationships and genes, all of which occur in great diversity among those species that we find around us.'

We also need to recognise that travelling frequently does not necessarily mean becoming more knowledgeable about birds or better birdwatchers. We need to ask why this idea is still so pervasive at a time when the call, song, image and published research about most bird species is so readily available at the touch of a screen. If we do not ask these questions, historians will. They will wonder why, in the midst of a climate crisis, prestige and status—inequality based on differences in esteem and respect—were so closely dependent on burning fossil fuels.

We also need to acknowledge that travelling far and frequently does not necessarily make us better or more cosmopolitan people. Along with information about birds and their habitats, we can access information about the people who call those landscapes home. We can learn about their cultures, economies, religions, political struggles and aspirations. We can find this information in newspapers, books, and specialist journals and magazines. We can watch mainstream and independent news and documentaries, enjoy good cinema and read excellent literature. We can talk to people from all corners of the world, including those who are at the forefront of environmental struggles, using communication technologies; or we can simply talk to those living in our communities who come from different cultural backgrounds. Obviously, direct experience of distant places, birds and peoples can provide wonderful insights, but if our genuine concern is learning about the world and being open-minded, these may not necessarily be the most relevant—at any rate, not always and especially not in a climate emergency.

Most importantly, though, we do not have to stop travelling, we just have to do it differently. As Ed Gillespie has noted, 'This is not the end of travel, but we are going to have to change how we do it, how far we go, for how long and how often. I think we are entering an era of far fewer, perhaps longer trips, with many more of them taken closer to home or overland.'[4] If your aspiration is to experience cultural and ecological diversity you can still do so. A whole continent with 24 official languages, 96 non-official languages, 70 ecoregions and more than 500 breeding bird species, is accessible on train (and this is without considering the large parts of the Maghreb which are also accessible by rail). As for visiting regions and continents beyond Europe, well, at the moment, and in the absence of more flexible schemes to organise work and school time and without more convenient low-carbon, long-distance transport systems, we will have to treat such trips as real and occasional luxuries.

More and more birdwatchers are embracing self-moderation in the way that they enjoy and study birds. The reasons for this are varied. First and foremost there is a growing understanding of the full and irreversible implications of climate change. This can have a paralysing effect but can also prompt individuals to action. There is also disillusionment with 'birding in the fast lane' and its emphasis on a hectic quest for the rare and the exotic by means of burning fossil fuels. Related to this, there is apathy towards narcissistic affirmations of self often found in the celebration of long national and world lists (and a sharpened awareness that a long list may simply reflect the privilege afforded by holding a certain passport and the financial means needed for such expensive pursuits). There are feelings of outrage at the inaction of politicians and the destructive actions of oil executives, and a desire not to be part of a system that is destroying conditions for civilised life on Earth by unnecessarily increasing fossil fuel demand. Advocates of low-carbon birding are unlikely to see themselves as innocent, passive victims and instead acknowledge that reducing demand is a key part of system change. There is also a growing realisation that wildlife tourism does not always benefit poor local people or wildlife. Finally, there is a growing recognition that our health is inextricably entwined with the health of the planet and that we can enjoy birds in ways that have positive effects on both.

We can all help to normalise moderation in the way we enjoy birds; and the greater our influence the greater could be our role in making this happen as fast as possible. Bird conservation organisations and the wildlife media in particular can play a key role. For decades they have communicated the idea that if people 'connect' with nature they will behave in more environmentally friendly ways. This is clearly not enough if the result is a desire to visit distant landscapes by means of burning more fossil fuels. We also need a message of sufficiency: we can live good lives within limits; we can enjoy nature without claiming a disproportionate share of resources; we can celebrate this unique planet that we call home, but it is a home we share with 7.9 billion people; we can limit ourselves collectively to allow all to enjoy this good life.

The message of sufficiency has to be accompanied by the truth about the climate emergency. Most birdwatchers and conservationists have shied away from speaking out about the scale of the problem and what this means for lifestyle change, sometimes fearing that they are going to scare people and lead them to despair and inaction. But this assumption is wrong.

Genevieve Guenther, expert on the role of language in the politics of the climate crisis, has explained it clearly: 'For many years the social science research was taken to show that scaring people about climate change was counterproductive, but in fact the original research argues that fear can be motivating when balanced by a sense of agency', that is, by showing people what they can do.[5] Conservation organisations and the wildlife media can help convey a sense of urgency by amplifying the voices of climate scientists who are explaining clearly what needs to be done, not what politicians and people with high-carbon lifestyles find palatable.

A liveable planet will be one in which we have replaced a culture of excess for the few with a culture of moderation and sufficiency for all. The aim of this book is to show that sufficiency in birdwatching is not only good for the planet but it can bring untold pleasures.

Structure of the book

The book is divided into two parts. The first part consists of two chapters about the need to rethink our birding habits. The first of these was first published in *British Birds* in April 2018 and is entitled 'Are we addicted to high-carbon ornithology?' Unexpectedly, this short piece captured a sentiment widely shared by many readers of the magazine and prompted soul searching in some quarters of the birding world. It summarises the need for a different birdwatching culture, one that is not wedded to the carbon economy, and it is therefore appropriate that it accompanies this collection of essays.

Due to limitations of space, in this *British Birds* article I could not address specific questions which birdwatchers concerned about climate change had been asking: Can we carry on driving as usual to watch birds if we switch to electric vehicles? If we fly less, what will happen to biodiversity-rich areas where conservation work is funded by wildlife tourism?

Can we 'cancel out' or 'balance' our emissions from flying and driving through forest preservation offsets? Can we use natural ecosystems to remove carbon so that we can carry on with our high-carbon lifestyles? Should we be focusing on collective action and system change rather than changing our lifestyles? I address these questions in Chapter 2, showing the gap between the dominant narrative that new technologies and gradual change are enough to address the climate crisis and what climate and sustainability scientists are actually telling us. The opinions expressed in this first part of the book are mine and not necessarily those of the authors of the individual chapters that follow.

The second part of the book consists of 29 short essays illustrating different ways of understanding and practising low-carbon birding. Each chapter can be read independently although I have grouped them around four broad themes: patch birding, birding holidays by train, personal reflections about embracing low-carbon birding, and personal accounts from professional ornithologists about their research on the way climate change affects birds. Collectively, these essays show that we can find ways to uncouple birdwatching from the carbon economy, and that our efforts to align our actions with climate science can go hand in hand with good birdwatching.

Before ending this introduction, a brief mention should be made of the Covid-19 pandemic. Most chapters in this book were written during the summer of 2021, with memories of the lockdowns still fresh and health concerns affecting travel decisions. This has influenced the content in two ways. First, there are fewer chapters about low-carbon birding trips and holidays than initially planned, as some contributors who had decided to write about their train journeys during that summer had to cancel or postpone their trips for obvious reasons. Second, the pandemic is the inevitable context in which some contributors reflected on and wrote about low-carbon birding.

In this respect, the book illustrates a pattern that will become more frequent in the coming years. Researchers studying lifestyle change have found that people are more likely to reconsider their assumptions and expectations about what is 'normal' in their everyday lives and make bigger decisions on the way they travel when their life circumstances alter due to, for example, a change of job or housing location, having children, an energy crisis or, as in this case, a public health crisis. Researchers have also found that some people use the opportunity provided by these changes to realign their attitudes and values with their actions.[6] This is what happened to some of the contributors. They were already concerned about climate change and had started to make changes in their birding, but the pandemic provided the opportunity to turn those incipient changes into habits. This is going to be a recurring pattern in the coming years as climate and energy crises intensify, the world faces new, perhaps more severe pandemics, and more and more people begin to grasp the realities of climate breakdown.[7]

1

Are We Addicted to High-Carbon Ornithology?

Javier Caletrío

Originally published in British Birds *111.4 (April 2018): 182–85.*

In ornithological company, the concept of climate change is usually discussed in terms of measurable impacts on bird populations and habitats, rarely on the way that it may affect the practice of birdwatching itself. This is remarkable considering quite how carbon-intensive certain styles of birdwatching have become. The days when Horace Alexander, one of the pioneers of British ornithology, expanded his usual New Year's Day birding walk around his home town with a train ride to Dungeness in 1910 or, after moving to the Midlands in 1918, a bus ride to the Lickey Hills, seem distant to generations of birdwatchers that have grown up taking the car and the aeroplane for granted. New Year's Day 2019 will find many birdwatchers driving to favoured spots or enjoying colourful birds in a warmer climate, perhaps around the

Mediterranean or even on a different continent altogether. It seems unlikely, however, that the coming decades will witness the continued growth of mobility that has characterised ornithology in recent decades. At least there are reasons to believe that it shouldn't.

Growing numbers of birders feel increasingly uneasy about their carbon footprint and there are recent references to the possibility that birdwatching in the future may involve travelling less or travelling differently. But the enthusiasm with which trips to distant places and long world bird lists are still celebrated suggest that, in general, British ornithology remains seemingly oblivious to the sense of urgency in recent debates about global warming. Climate scientists argue that what matters is not so much levels of technological efficiency and emissions reductions in a more or less distant future, but *cumulative* greenhouse emissions, which could trigger a tipping point in climate dynamics with potentially devastating consequences.[1] The cumulative emissions concept reframes global warming as a short-term issue, in which action taken in the next two decades could be critical. During this period, unprecedented levels of emissions reduction are necessary, and the longer we delay these cuts the more abrupt reductions will have to be in the future.[2] The key issue here is that since the time required for adapting everyday infrastructures to new energy systems is usually measured in decades, there is no mathematical alternative but to reduce energy demand if there is a reasonable chance of avoiding 'dangerous' climate change. This means that changes in lifestyle are unavoidable. In terms of transport this means shifting to cleaner options such as railways and cycling and, especially for those who have lifestyles with large carbon footprints, flying less or stopping flying altogether.[3]

Initiating the debate about achieving a significant de-carbonisation of ornithology is not easy. Routine short- and long-distance travel underpin contemporary economic and social life and there are strong economic and cultural inertias. Conservation organisations like the Royal Society for the Protection of Birds and the Wildfowl and Wetlands Trust are part of the tourist economy and rely on a membership that places a high value on access to parts of the countryside that are poorly served by public transport. Similarly, over the last few decades, with the desire to watch birds abroad, there has been a proliferation of specialised travel agents covering every continent. A reduction of 'birdwatching miles' would affect these organisations and companies.

Questioning high-carbon travel may prove even more challenging culturally. Celebrated figures in European ornithology, both dead and alive, are generally well-travelled individuals, and a brief perusal of birdwatching literature shows that travel is an essential part of what it means to be an accomplished ornithologist today. To many people, birdwatching in distant places is an expression of a curious and caring disposition towards the world and its birds, and while this can be experienced as a deep personal feeling, it also entails a broader social dimension. In some circles, visiting exotic destinations and having long bird lists are regarded as signs of distinction and expertise.

Addressing the problem of high-carbon travel is not simply a matter of educating people. One of the paradoxes of the current environmental crisis is that educated, environmentally aware segments of the population are often among the highest carbon emitters.[4] In Germany, Green Party supporters fly more frequently than supporters of any other party.[5] Research in the UK and Norway shows that people are happy to adopt environmentally friendly behaviour at home but are reluctant to give up their holidays abroad. Some even regard those trips abroad as a time when environmentally friendly behaviour can be suspended as a 'well-deserved treat' for their commitment to the environment at home.[6] This stark dissonance between local aesthetics (for example, buying organic and/or local food) and the global picture illustrates the challenges of promoting lifestyle change when travel has become so engrained in contemporary ways of life and touches deeply on our identity.

Despite these difficulties, it would be a mistake to frame the transition towards low-carbon ornithology in terms of sacrifice or as limiting choice. A debate on the way that climate change may redefine birdwatching should underline the opportunities that a reduction in travel would open up. We should also acknowledge that British birdwatching encompasses an extremely diverse set of styles and practices, some of which are already low carbon. An obvious example is local patch birding. This style of birding, revolving around the enjoyment of gaining an intimate knowledge of a particular place and its birds through long-term engagement, should be celebrated and encouraged. *British Birds*'s recent editorial on the SK58 Birders and the new section 'My patch' are timely examples of the kind of institutionalised encouragement needed for cultural change in the world of ornithology.[7] What's more, this approach has the potential to enhance local knowledge of population dynamics, to boost participation in national and international monitoring schemes, and to engage with the local population.

Importantly, low-carbon birdwatching does not necessarily involve giving up trips abroad. Many destinations can reasonably be reached by train and/or ship. Travelling from England to the northwest of the Mediterranean can be done in a single day by train, often with less hassle, more comfortable seats and better views than flying. Similarly, low-carbon ornithology should not mean giving up birding as a competitive 'sport'. Traditionally, that has involved an excess of birdwatching miles but low-carbon lists could be encouraged instead as a more genuine sign of distinction. The growing popularity of initiatives such as Patchwork Challenge,[8] and particularly their Green List category, is evidence of the potential appeal of redrawing your birding horizons.

While it is impossible to predict how a transition to low-carbon ornithology might unfold, it is possible to identify some of the key issues. One of these is the rise of carbon inequality. According to recent research, the richest 10% of the world's population is responsible for around 50% of carbon emissions.[9] This inequality is even starker when regarding transport-related emissions. In France, 5% of the population is responsible for 50% of transport-related carbon emissions in tourism, mostly as a result of flying (and 20% of the population is responsible for 80% of emissions).[10] In the UK, just 15% of people take 70% of flights, most of which are leisure flights by frequent flyers.[11] To my knowledge, there is no research on the mobility patterns and carbon footprint of British birdwatchers, but it is reasonable to assume that a significant proportion of dedicated birdwatchers fly more frequently than people in the same income bracket and that the largest carbon footprints correspond with those with higher incomes. These trends are likely to be reinforced in the immediate future as inequality is set to continue rising in an anticipated context of low economic growth and weak political support for progressive taxes.[12] As this happens at a time of mounting pressure to act on climate change, we may expect debates about transport-related carbon footprints to redefine what is regarded as a legitimate and ethical way of enjoying and studying birds, and perhaps even question the late twentieth-century image of birdwatching as an expression of more democratic and equal societies.

By now some readers may be wondering whether a debate on the transition towards low-carbon ornithology is even necessary, considering the negligible effect of British ornithology on the climate system. I would argue that it does matter, for two main reasons. Firstly, regarding distant travel, while aviation's contribution to human-induced climate change is currently smaller than that of other sectors (4% of observed human-induced global warming to date),[13] the problem is the rapid growth in the share of emissions. International aviation's CO_2 emissions may represent 22% of global emissions by 2050.[14] This share is greater in countries where aviation is more prominent. Projections for the UK show that if the government is committed to limiting global warming to 1.5°C and yet gives the green light to the planned Heathrow expansion, 71% of the national emissions budget will be consumed by aviation by 2050.[15] To put things in perspective, bear in mind that less than 10% of the world's population has ever been on a plane.[16] Those of us who fly in order to watch birds are part of a small, highly polluting elite. The obvious question here is that if in the face of this evidence we still feel entitled to fly frequently, why should middle classes in emerging economies not feel so too? Currently, 423 new airports are planned or under construction, with 223 of these in the Asia-Pacific region alone.[17] At the moment, there are no convincing signs that a major breakthrough could help to reduce aviation's emissions substantially, and the limited opportunities available to improve fuel efficiency cannot compensate for the expected growth in demand.[18] This growth in demand can happen within the available carbon budget only if other sectors curb their carbon emissions even further. Climate scientist Kevin Anderson argues that 'expanding aviation is numerically, technically and symbolically incompatible with commitments to avoiding dangerous climate change.'[19]

Secondly, and crucially, this debate matters because the credibility of our claims is at stake. As a collective that professes a concern for the state of the planet, at least for those who accept the scientific evidence of human-induced climate change, our actions as ornithologists cannot imply a tacit denial of that evidence. In advocating the vision of a world informed by science, the way we communicate that science cannot simply be about accessible writing styles or reaching the wider public through different media. It is also about nurturing trust between the public and ourselves. And for our message to be trusted, our action needs to be seen as being consistent with the message that real and urgent action is needed on emissions. This is relevant not just to individuals and organisations with a higher public profile. To the extent that British ornithology is admired and inspires individuals and organisations around the world, we all have a responsibility in making that transition towards low-carbon ornithology a successful one. The same urgency we demand in addressing the plight of endangered species and habitats should inform efforts to change our own carbon-intensive lifestyles.

We need a debate about how best to encourage a transition in British ornithology because the path is not easy and there are tensions and interests that need to be balanced. Vital research and conservation work will still rely to some extent on oil-fuelled transport. Some groups in society such as the elderly and those with disabilities rely on cars for basic access to the countryside and cannot always make use of public transport. Nature reserves will still need to generate income from visitors. And all of us will require role models. A transition to low-carbon ornithology will need its own heroes but their profile will be different from some of those celebrated in recent years. Surely some of these unsung heroes already exist. They are unlikely to care too much about long world lists and are more likely to be found on their local patch, with years of consistent commitment to understanding the birds around them. Other heroes will emerge from younger generations eager to explore distant places without costing us the planet. A cultural change on this scale will succeed if those willing to commit find support among friends and local birdwatching clubs, and if these efforts are given the institutional support and visibility that they deserve.

I doubt that Horace Alexander's experience of birdwatching in the early twentieth century was any poorer than ours simply because there was less fast travel or bird lists were shorter. In reading about the experiences of pioneers of twentieth-century British ornithology, I cannot avoid wondering whether we may have been the victims of the tyranny of choice in these times of fast living. Maybe it is time to enjoy not just slow food, but also 'slow ornithology'.

2

Questions of Travel, Climate and Responsibility

Javier Caletrío

Technological innovation is often invoked to justify making little if any change to the high-carbon lifestyles of affluent people. But the science and the maths of climate change, the timeframe for tackling it and broader sustainability issues do not really leave us with much choice but to reduce consumption and to ration carbon emissions. In this chapter I address this issue by focusing on the role we can play as citizens, consumers and role models, the need to consider the question of equity when discussing how much we need to reduce driving and flying, and how to broaden the conversation about the future of wildlife tourism.

Why do my actions matter? Is a focus on lifestyle change a distraction from addressing system change?

Rapid and significant emission reductions require decisive action by governments and big business to put in place regulations and infrastructures that will enable individuals to change their habits and live more sustainably. But the argument that a focus on lifestyle change is a distraction fails to acknowledge that system change is as much about demand as it is about supply,[1] and that the actions of individuals can be connected to larger efforts for systemic change.[2]

Adopting climate-friendly habits such as following a plant-based diet, driving less and flying less matters because each change can be a catalyst for collective action.[3] If, for example, you decide to fly less and you talk about it, you are inspiring others. People are more likely to adopt climate-friendly habits when others around them, especially influential figures, do so.[4] Recent research has shown that of those who know an individual who has decided to fly less, around half fly less themselves as a result, and around three quarters say knowing that person has changed their attitudes.[5] When communicated effectively, the action of an individual sends ripples across the many social relations that each of us is part of—local communities, workplaces, professional associations and leisure societies. Individuals collectively matter, as we influence each other and our actions can help shift norms and make major political actions more likely.[6]

A focus on the demand side of system change also helps to shed light on the role of the high emitters in the climate crisis. For most of the population, major cuts in CO_2 will be via structural change across the energy system. But for a minority of high emitters, lifestyle change is the most effective way to reduce emissions. The top 10% of earners globally, which constitutes individuals with an annual income over £27,000, are responsible for half of all carbon emissions through their individual consumption habits.[7] The emissions of this globally affluent minority alone are on track to consume the carbon budget for 1.5°C by 2030, irrespective of what the rest of the world does.[8] In the UK, consumption emissions of the richest 10% in 2030 are set to remain five times above the global average per capita level that is compatible with 1.5°C.

The world's richest 10% consumes around 20 times more energy than the poorest 10%, and a large portion of this consumption by more affluent people is for travel: flights, holidays and big cars driven long distances.[9] In the European Union, land travel (purchase of vehicles, transport fuels and services) represents 32% of the carbon footprint of the richest 10%.[10] Although we all live within systemic constraints, it is clear that there are degrees of responsibility. If this top 10% of emitters were required to cut their carbon footprint to the same level as the EU average, that alone would deliver around a 30% cut in global CO_2, significantly increasing the chances of staying below the 1.5°C target.[11] If we were serious about the climate emergency, policies could be driving such cuts almost overnight.

Why do we have to drive less?

Transport was responsible for 27% of greenhouse gas emissions in 2019, making it the UK's largest source of emissions.[12] More than half of those emissions come from cars. The problem is that although fuel efficiency has increased over the last decade, so too have miles driven, and the result is that emissions are not decreasing. Reducing emissions from the transport sector by 2030 requires reducing absolute levels of miles driven.[13]

So how much do we need to reduce driving? This depends on the ambition of our mitigation targets and whether we respect the principle of equity enshrined in the Paris Agreement—that all countries must contribute to climate action, but those who have emitted most historically have a greater responsibility.

Honouring the temperature and equity commitments in the Paris Agreement demands that wealthy industrial nations meet more stringent mitigation targets than politicians are willing to accept.[14] What does this mean for the car sector? According to Kevin Anderson, professor of energy and climate change and former director of the Tyndall Centre for Climate Change Research, even with a complete transition to electric cars and clean electricity by 2035, car use (vehicle-kilometres) needs to be reduced by 40–60% if the sector is to make its fair contribution to delivering on the Paris commitments.[15]

Equity needs to be at the core of any meaningful mitigation policy, and this applies not just between countries but also within countries: between wealthier and poorer parts of the population. The responsibility for these reductions in miles driven should be distributed fairly, with greater effort being made by those who drive more for leisure. In the UK, 24% of households do not have a car and those households with the highest incomes use their cars three times more than those with the lowest. There are also spatial inequalities in the ability to reduce car use. Generally, residents in urban areas have better access to public transport than those in rural areas. This also means that conservation and monitoring projects remain more car dependent in rural and remote areas. Even in a context where carbon needs to be rationed, it is important to guarantee the continuity of this vital work.

An important fact to bear in mind when reassessing our travel habits is that in England only 3% of trips by car are over 50 miles, but these represent 30% of the total mileage, so reducing unnecessary long-distance car journeys matters.[16]

Why do we need to fly less?

The number of air passengers worldwide increased from 1 billion in 1990 (the year when the International Panel on Climate Change issued its first report) to 4.5 billion in 2019. In 2019 the International Civil Aviation Organization estimated that this figure would increase to 10 billion by 2040. Despite claims by the aviation industry about rapid progress in decarbonising flying, such growth in demand is not 'green'—at the moment, there is no such thing as sustainable commercial aviation.[17]

Innovations in fuel efficiency and less polluting fuels are not enough to compensate for the current and expected rapid growth in demand. Between 2013 and 2019 passenger air traffic increased nearly four times faster than improvements in fuel efficiency.[18] Regarding 'sustainable aviation fuel', no credible analysis expects all future jet fuel to come from sustainable sources by 2050. The diminishing ambition of the targets for sustainable jet fuel set by the Internal Air Travel Association (IATA) over the last decade illustrates the gap between rhetoric and reality. The target in 2009 was 10% by 2017; in 2011 it was 6% by 2020; and in 2020 it was 2% by 2025. Actual sustainable fuel use in 2018 made up just 0.002% of airline fuel use (enough to power about ten minutes of global air travel).[19]

The sustainability of alternative fuels has also been questioned. Production of aviation fuels from biomass at scale poses threats to biodiversity to and farmers' livelihoods,[20] and diverts support from other renewable energy sources. But even if biofuels were sustainable, they should not be used for unnecessary consumption, such as the holidays of a small minority of frequent flyers, until all other facets of society such as health and education are zero carbon. In the next two decades scarce, zero-carbon energy sources should be allocated to socially relevant priorities.

Synthetic fuels are also being considered for decarbonising fuel demand, but these require enormous amounts of electricity. The organisation Transport & Environment argues that 'using electrofuels to meet expected remaining fuel demand for aviation in 2050 would require renewable electricity equivalent to some 28% of Europe's total electricity generation in 2015 or 95% of the electricity currently generated using renewables in Europe'.[21]

Electric aeroplanes are not the solution either. Although the aviation industry heralds the advent of electric planes, engineers acknowledge that the roadmap for developing this technology is a long one. Small electric planes for local and regional flights may operate within the next two decades, but there is no prospect yet for commercial short- and long-haul flights, which in the UK are responsible for 87% of aviation emissions.[22] Currently, no company is trying to develop a battery plane in order to fly long distance. Technologies will keep developing, but at the moment there is no feasible replacement for fossil fuel aviation at its current scale anywhere on the horizon—certainly not by 2050.

To sum up, the potential of alternative fuels to decarbonise aviation is limited and commercial electric flights are still a distant dream. Right now, the only way to bring down emissions significantly in aviation within the next decade is by reducing demand.

So, by how much should aviation demand be reduced within a decade if we take the Paris Agreement temperature commitments at face value—that is, without relying on tentative technologies to remove carbon from the atmosphere? As I explain later in this chapter, these technologies exist only as prototypes and may not work at scale.[23] According to the Centre for Alternative Technology, without considering carbon-removal technologies, demand for fossil fuel aviation in the UK needs to be reduced to a third of pre-pandemic levels by 2030; and according to FIRES UK (a collaboration between the universities of Cambridge, Oxford, Nottingham, Bath and Imperial College London), it needs to be halved by 2030, aiming towards zero by 2050. For now, complying with the Paris Agreement requires reducing aviation demand with policies targeted at frequent flyers and by changing travel norms.[24]

Aviation tends to dominate the carbon footprint of frequent flyers and this is not likely to change in the coming decades. Considering planned fuel and operational efficiency gains, the climate impact of a return long-haul flight from London to Jakarta, for example, will still be in the range of 2–3 tonnes of CO_2 by 2050. Yet, globally, emissions per capita need to go down to 2.5 tonnes in 2030, 1.4 tonnes in 2040 and 0.7 tonnes in 2050 if we are to limit warming to 1.5°C. Emissions per capita in the UK in 2019 were 8.5 tonnes.[25]

Can we 'cancel out' or 'balance' our emissions from flying through forest preservation offsets?

The problem is that CO_2 stays in the atmosphere for a long time (between 300 and 1,000 years) while the trees bought or planted through an offsetting scheme may not be there next month, next year, or in the next decade. Fires, changes in governments, political instability, poverty, corruption, revolutions, armed conflicts, migrations, 'natural' disasters—it is impossible to predict what will happen to a forest (let alone to the company selling the carbon credits) in the future, and there is little reason to believe that the coming decades are going to be any less troubled than the last century.[26]

According to Bonnie Waring, ecologist at Imperial College London, 'the fact is that there aren't enough trees to offset society's carbon emissions—and there never will be'. The potential of forests to absorb carbon, she argues, is limited: 'If we absolutely maximised the amount of vegetation all land on Earth could hold, we would sequester enough carbon to offset about ten years of greenhouse gas emissions at current rates' and it would take them about a century to grow large enough to do so. 'After that, there could be no further increase in carbon capture.' Waring concludes that 'land ecosystems will never be able to absorb the quantity of carbon released by fossil fuel burning. Rather than be lulled into false complacency by tree planting schemes, we need to cut off emissions at their source.'[27]

We need forest conservation, reforestation and potentially afforestation in addition to, not as a substitute for, virtually eliminating emissions.

Can we remove carbon from the atmosphere with carbon-removal technologies such as Bioenergy with Carbon Capture and Storage (BECCS)?

Removing carbon from the atmosphere is a complicated process initially conceived as a backup tool in case emissions are not curbed as early as needed and to deal with sectors that are difficult to decarbonise, such as agriculture. Climate modellers, however, have incorporated NETs in their models as if they already existed and were capable of removing vast amounts of carbon. The reality is that these technologies exist at best as small pilot schemes and may never work at the required scale. The European Academies' Science Advisory Council clearly noted in 2018 that 'Negative emission technologies may have a useful role to play but, on the basis of current information, not at the levels required to compensate for inadequate mitigation measures. ... We conclude that these technologies offer only limited realistic potential to remove carbon from the atmosphere and not at the scale envisaged in some climate scenarios'.[28] The tentative potential of NETs is being used to undermine the requirement for immediate and widespread decarbonisation.[29] An additional problem is that the deployment of BECCS would require between 0.4 and 1.2 billion hectares of land (that is the equivalent of 25% to 80% of all land under cultivation or the equivalent of up to three times the size of India), which could be achieved only by reducing land available for natural habitats and food production.[30]

Can we use natural ecosystems to remove carbon? If we invest in 'nature-based climate solutions', can we carry on with our high-carbon lifestyles?

Around one third of the greenhouse gas mitigation required between now and 2030 can be provided through ecological restoration, a set of processes known as 'Natural Climate

Solutions'.[31] The widespread adoption of cost-effective, natural, negative-emissions approaches such as reforestation and wetland regeneration along with stringent mitigation measures (including lifestyle change) can help deliver the Paris Agreement and reduce or even eliminate the need for BECCS. The regeneration of natural landscapes is therefore vital for a stable climate, but this is not a substitute for urgent and drastic emission cuts in line with climate science.

Will some biodiversity-rich places disappear without wildlife tourism?

Wildlife tourism can have positive effects in some places, for some people and some species. This is a good starting point for a conversation about the sustainability of wildlife tourism. In broadening the conversation we also need to consider whether all wildlife holidays have conservation value (unfortunately, the response is that often it is negligible or non-existent). For those places where tourism does provide vital funding for conservation, we must ask whether there is an overreliance on international tourism, what led to this situation, and whether it is sensible to continue relying on tourism to fund conservation in the face of a growing risk of pandemics and the need to reduce aviation demand within the next two decades.

In examining why some places have become dependent on international tourists one needs to consider the wider economic and institutional context of programmes to develop tourism as a conservation tool. Since the 1990s, international aid for development has often been tied to structural adjustment policies that have reduced the size of states, resulting in a lack of state capacity to enforce conservation laws and the need for foreign currency.[32] This development approach envisions market mechanisms such as ecotourism, bioprospecting, and carbon offsetting as providing key funding for biodiversity conservation. A legitimate question is why should biodiversity conservation and the future of a liveable planet ultimately depend on the market?

If international tourism is the main source of income in a locality, we need to ask why this is so and what the implications for people and the environment are. Is this specialisation the result of conservation policies limiting or prohibiting permanent human settlements and other economic activities? If so, who made this decision and who benefits from it? Did the local population have a voice or was it imposed by national governments, conservation organisations, local elites, tour operators and international development agencies? Were conservation policies informed by the values and worldviews of the local population? Were human rights and land rights violated?[33]

The creation and management of protected areas such as national parks have often been inspired by a 'fortress conservation' approach, which on the one hand limits what can be done and who can live within protected areas, and on the other seeks to bring largely affluent tourists into contact with nature. Fortress conservation is the dominant approach of conservation globally and has led to human rights violations. According to political ecologist Daniel Brockington, by 2002 the expansion of the global network of protected areas had resulted in the eviction of a minimum of 10 million people from their homes and lands.[34]

There are, however, other biodiversity conservation approaches inspired by principles of equity and environmental justice which promote structural transformation of the economy aimed at greater resilience in the face of economic and ecological uncertainties.[35] Advocates of these approaches have proposed a 'conservation basic income' that sustains biodiversity-friendly livelihoods.[36]

To summarise, policies towards deregulation, forcing the opening of national markets to trade and capital, and the shrinking of governments via austerity or privatisation constitute the context in which 'conservation-funded-through-market-mechanisms' in biodiversity-rich areas emerged during the last four decades. But other conservation paradigms are possible.[37] Flying-dependent tourism from affluent countries may not have to be the only economic option in biodiversity-rich areas.

The conversation about the sustainability of wildlife tourism also needs to consider whether it is a viable conservation tool in the face of a growing risk of pandemics. Since the 1990s scientists have warned about the possibility of health crises like Covid-19 in a context of rampant ecological destruction, intensification of farming, growing urbanisation and increased international connectivity, among other processes. The likelihood of more frequent and severe pandemics is only increasing. Since the turn of the millennium, we have had six significant threats: SARS, MERS, Ebola, avian influenza, swine flu and now coronavirus. Notwithstanding the tragic loss of many lives globally, Covid-19 could have been even worse. But what about the next pandemic?[38]

The pandemic has demonstrated the fragility of the 'conservation-funded-through-tourism' model. An exclusive focus on international tourists and the neglect of domestic tourism and other activities has proved to be detrimental for people and wildlife.[39] Perhaps, rather than insisting on the fact that some areas depend on tourism to justify business as usual, we should shift the conversation towards discussing ways to lessen dependence on long-haul flying and examine other conservation paradigms and sources of income.

Am I claiming that we should stop wildlife tourism altogether? No, I am not arguing that we should put an end to travel and tourism in natural areas. Rather, I am arguing that we should be agnostic rather than enthusiastic about wildlife tourism, especially if it involves high-carbon transport. The tourist industry needs to cut emissions in line with climate science and focus on markets that require no flying or less flying. This is critical for the future of any tourist destination. With no technological breakthrough in sight to make aviation a clean mode of transport within a timeframe compatible with climate change targets, and with growing social pressure from the climate movement to curb demand, it is in the interest of destinations to become less reliant on international tourists by focusing more on markets accessible by less-polluting modes of transport and considering other land uses compatible with the preservation of nature. How to do this in a just manner should be the object of an open conversation about how places, companies and organisations that rely on growing levels of flying can adapt to the realities of the climate emergency. Central to this conversation should be the question of how to share the carbon budget available for aviation equitably and for what purposes flying is deemed essential.

3

The Seven Cs of Patch Birding

Nick Moran

It took almost 40 years for the habit of birdwatching in a local area, 'patch birding', to become truly embedded as my primary birding activity. Spending my formative years in suburbia in the Vale of York, in a small town surrounded by arable fields, I relied on my parents to drive me to the Lower Derwent Valley (25 minutes), Fairburn Ings (45 minutes) or even further afield to the likes of Blacktoft Sands and the Yorkshire coast. Even the eight-mile return cycle ride to Strensall Common, one of the few remaining lowland heaths in Yorkshire, was something I rarely did. With hindsight, four years spent studying in Fife during the mid 1990s started my glacial-paced transition to a form of birdwatching less reliant on the internal combustion engine. Looking round various universities, it was immediately apparent that a key advantage of St Andrews over others (even Norwich!) was the wealth of birdlife on the doorstep, including a host of species that were unfamiliar to me. Back then though, full of the new-found freedom of student life, the net for what comprised my patch was cast rather wide! Pedal-powered excursions took me all over the East Neuk of Fife (Largo Bay, Fife Ness and along the coast between there and St Andrews) and northwards to the Eden estuary, Tentsmuir and as far as the outer reaches of the River Tay. While this more than satisfied a rapidly expanding appetite for birding exploration and provided many memorable experiences, my approach to choosing a destination for any given trip was rather scattergun. As a result, I did not develop a particularly strong association with, or understanding of, any particular spot.

Three overseas teaching positions during the 2000s broadened my horizons but, conversely, also began to focus my local birding on more clearly defined patches. In each of southwest Uganda, Shanghai, China and Abu Dhabi, United Arab Emirates (UAE), either not owning a car or the experience of living in a large city on a major migration route meant more of my birding effort was concentrated on a few relatively small sites close to home. In Shanghai and Abu Dhabi the city parks were often rewarding birding venues, particularly in spring and autumn. In both cities I adopted a couple of green areas that were within walking distance of home as my local patch. However, in neither case was there a conscious decision to define and study a patch; instead they were born out of necessity, particularly in Shanghai.

Birding was still relatively uncommon in China in the early 2000s and exploring the potential of the city parks had a sense of the pioneering about it. By the time I was considering moving to the UAE though, the global birding scene was connected enough for me to find and contact an Abu Dhabi-based birder to discuss the prospects for local birdwatching. Learning about the wealth of exciting local birding opportunities played an important part in the decision to move there. However, it would be disingenuous to imply that local birdlife was the main birding motivation for working overseas—I did travel extensively to watch birds both within and beyond each country during that period.

Driving and flying primarily or exclusively for birding was central to my birding behaviour during the 1990s and 2000s. By the late 2000s though, cognitive dissonance around the conflict between concern for the environment and how I was enjoying wildlife started giving way to a growing sense that high-*carbon* consumption, particularly for a nature-based leisure pursuit, was unsustainable for the planet and for my own state of mind. Placing a much greater emphasis on birding an area near home has been instrumental in helping me to modify that behaviour. Of course, if somewhere is to be a practical option as a patch—a place that can be visited regularly and whenever a brief window of opportunity opens— then by definition, it must be local. If it can be reached on foot or by non-motorised or public transport, better still. Ideally, it will also have sufficient diversity and potential to maintain the interest, more of which later. After my wife and I decided to return to the UK, I was lucky enough to be offered a job with the British Trust for Ornithology (BTO) and, within a few months of starting, be able to move to a property overlooking the BTO's Nunnery Lakes reserve. While good fortune played a role, local birding possibilities were an important part of the overall decision to move to East Anglia and the specific location in which we chose to live. The status of Norfolk as a birding Mecca is well known and I had been visiting since my teens. Buying a house adjacent to the Nunnery Lakes guaranteed a relatively high diversity of birds and other wildlife on our doorstep, not to mention being within very easy walking distance of the BTO's offices.

Regularly birdwatching a site and adopting it as your own provides unlimited scope for discovery of and connection to the place. One appealing aspect of patch birding that has crept up on me but that I now find particularly motivating is the *continuity*. A few local birdwatchers were keeping an eye on the Nunnery Lakes well before the BTO moved its headquarters to Thetford in 1991, and several current staff members visit regularly. Bird records for the site stretch back to the early 1980s and there have been monthly waterbird counts for the Wetland Bird Survey (WeBS) since 1991. Contributing to a sequence of recording that now stretches over five decades is both rewarding and revealing. During nearly all my visits, I use the BirdTrack smartphone app to log a 'complete list': a record of every species that I positively identified, rather than just the highlights. Maintaining this discipline provides a measure of absence (or at least, absence of detection) as well as presence. This in turn makes it possible to predict the likelihood of detecting a particular species at any point in the year, and to see how this has changed for each species between years and over my 12 years of watching the patch. Little egret, for example, was breeding within ten miles of the Nunnery Lakes by 2006 but featured on just 6% of visits in my first three years, 2009–11. In contrast, I recorded little egret on 73% of visits during 2019–21. Counting individuals of certain species has also been revealing: chiffchaff is a case in point, with a distinct peak in the last week of August, probably associated with local post-breeding dispersal. A similar approach to local bird recording in the UAE led to a paper on the timing of spring migration through Abu Dhabi.[1]

Consistently watching and recording at a patch yields fine-resolution information about the birds and other wildlife that use it. One fascinating consequence of this is the ability to place local records in the *context* of what is happening at regional, national and even international scales. This can take many forms, such as changes in abundance, distribution and the phenology of migratory species, as well as irruptions, hard-weather movements and other patterns of occurrence linked to the conditions at the time. For instance, a month after the unprecedented arrival of Siberian accentors and other eastern vagrants in Britain in October 2016, the Nunnery Lakes hosted its first little bunting and Siberian chiffchaff. Another contextual aspect is the mental adjustment that takes place as an understanding develops of what is and is not unusual at a certain site, and how this relates to other local spots. To illustrate this, 78 of my 112 UK records (70%) of jack snipe—a relatively uncommon and usually elusive winter visitor—are from my patch while, in contrast, just eight of my 771 UK records (1%) of redshank—a species that breeds ten miles away—are from the Nunnery Lakes. Raising a toast to a sighting of a species like redshank is a sure sign that the recalibration is complete!

Nature is under immense pressure from a multitude of human activities. In addition to lowering the individual birdwatcher's carbon footprint by reducing travel, building a deep knowledge of the natural history of a local area can be influential in shaping *conservation* measures. A long, detailed data set often plays a key role in informing decisions about land use change, the siting of new habitat features and the management of existing ones, among many other applications. This emphasises the value of a 'complete list'-style approach to documenting observations of birds: it is extremely hard to predict which species might be lost from or colonise a patch and, in turn, the assemblage of species it will be important for in future, hence the importance of recording all species encountered. Structured surveys are another powerful tool and though survey visits can be time-consuming, the standardised methodology delivers robust results. At the Nunnery Lakes, the same territory-mapping protocol that was used in the Common Bird Census was applied in the early 1990s and again since 2015, in tandem with nest-recording. Striking differences are apparent between the two periods, despite the relatively short gap: in 1991 there were 30 willow warbler territories compared to just three of chiffchaff but by 2018 the situation had reversed, with seven willow warbler and 23 chiffchaff territories. This pattern has continued and although nest-recording efforts were hampered by the lockdowns in 2020 and 2021, it is doubtful whether there were any willow warbler breeding attempts in the last couple of years. Accurately documenting such changes puts us in a better position to address them, or at least to refocus efforts on upcoming species assemblages. Furthermore, watching a patch can add fine-scale data for poorly known areas. Observations from a seemingly unremarkable inland patch may actually provide more insights than a well-watched section of the coast: inspiration for the many UK birders who live some distance from the sea.

Birdwatching can be a solitary activity, and solitude does have its benefits. However, many birders enjoy the sense of camaraderie associated with being part of a *community*. A dedicated patch-watcher's chosen patch is often shared with other people, some being birdwatchers of varying levels of experience and identification skill, others being non-birders who visit for leisure pursuits that are not directly linked to wildlife. Passing on personal observations can increase the positive feeling of belonging to a community, and perhaps even educate and inspire those who are less cognisant of their surroundings. A local birdwatcher's own comprehension of a site's natural history can be augmented by discovering what others have noted. Wider communities exist too, in both time and space. The long tradition of bird recording in the UK can give patch birdwatchers a sense of being part of an effort that began long ago, a history worth remembering. In the social media era, it is easy to feel part of a UK-wide and even global kinship of patch-watchers and for those motivated by friendly competition, both with others and themselves, there are volunteer-run ventures such as Patchwork Challenge.[2]

Memories of unusual discoveries or special experiences, interpersonal relationships developed over time and the emotional attachment to and feeling of ownership of a place all add to a sense of connection. Personally, this *compelling* aspect of local birdwatching manifests itself in all these ways and more: the shared discovery of a little bunting, the reading of the digits on a German-ringed cormorant over several days' observation, the deepening understanding of the timing of movements and changes in detectability of many common species, and the *esprit de corps* among the dedicated regulars. There is a sense of connection with, and an intimate knowledge of, some individual birds, too. BTO's cuckoo tracking project has involved the satellite-tagging of two male cuckoos at the Nunnery Lakes and it was captivating and enlightening to follow their progress and in cuckoo Valentine's case, see him safely returning the following year.

Whatever it is that inspires the individual birdwatcher to concentrate on a local patch—the satisfaction gained from long-term monitoring, the discoveries of locally rare species, the fascinating insights made possible by sound-recording nocturnal flight calls ('nocmig')—there is much to *celebrate* about birding locally! Encouraging more birdwatchers to adopt and focus on a patch near home is not only a rewarding element of the activity but makes a small but valuable contribution towards changing birdwatching behaviours for the better.

4

Understanding Our Local Birds

Angela Turner

I am captivated by the lives of birds, especially hirundines, lifelong favourites of mine. I can watch them for hours, losing myself in the here and now, totally absorbed in their behaviour. They make me feel calm. Nature therapy—the restorative effect of being in nature on our mental and physical wellbeing—has been much in the news, especially during recent Covid-19 lockdowns, whether that is forest bathing the Japanese way or just walking in a local park. I find observing the behaviour of birds greatly enhances the experience. Even the commonplace such as a barn swallow (henceforth 'swallow') babbling on a wire above me can bring a feeling of awe and positivity. Improving our own knowledge and understanding of the behaviour of birds increases the fascination and enjoyment of watching them. Their behaviours are often more nuanced than a cursory look may suggest. The swallow's 'tsi-wit' call, for instance, is easily recognised as an alarm call given when a predator is around, but you may also hear a male on his territory uttering it if another male comes too close. Swallows have other, less frequently heard, calls too: if you hear a low-pitched, slurred, two-note call, look for a hobby or peregrine close by as this is a warning of imminent danger.

Observing even common birds regularly in your local area reveals subtle and interesting behaviours and adaptations to local conditions. Swallows, for instance, are well known for catching insects in flight. Watch them on a chilly spring morning, however, and you might see them perching on a wall and pecking at a spider in a web, or in the evening they may be fluttering around a lamp to catch the moths spiralling in. One of my memorable experiences with swallows was on a hot June day on a farm in the Carse of Lecropt, near Stirling in Scotland, when they took advantage of a localised but abundant source of food. I was walking along the edge of a field when I saw a swallow leave a nearby stone barn and speed low over the long grass towards a hedge. I saw it turn at an oak tree where it started to skim round the branches, circling up the canopy higher and higher, snapping at some prey I could not see. I watched it for half a minute or so before it dropped down to grass level and headed back to its chicks. I stayed to see it come back again and again, as if intent on denuding the oak of its rich insect life. When I had a closer look at the tree, I was intrigued to see small green caterpillars, some dangling on almost invisible silk strands from the leaves they had been eating. I later found out that this pair of swallows had indeed been catching these caterpillars, which were the larvae of a moth, the white-shouldered smudge. Caterpillars had rarely been noted on the swallow's menu before. Hanging by a thread is a defensive behaviour which works well when predators approach on the leaf but is a fatal strategy against aerial insectivores.

Birds, however well known, continually amaze and surprise us, as browsing through the notes section of a journal like *British Birds* shows. For example, in June 2015 one birder on his regular patch at Rainton Meadows Nature Reserve near Durham, UK, saw a great crested grebe, normally a fish-eater, catching hirundines.[1] The grebe had a sand martin in its bill when first seen and was dunking it in the water. After swallowing it, the grebe caught two more, though the third managed to free itself. In the following two years at the same site a great crested grebe was seen to catch and swallow both sand martins and swallows. On each occasion harsh weather—rain, fog or strong wind—had forced the hirundines to feed low down over the water, where they were easy prey for the grebe.

Not all behaviour is so remarkable but is still worth recording. Notes in *British Birds* often add to our knowledge of birds' diets or nesting behaviours. One described a flock of 10 to 12 juvenile and adult house martins perching in an elder bush and feeding on the berries.[2] This is a fascinating observation; while occasional feeding on plant material is known among hirundines, and house martins have been recorded eating hawthorn berries, it is rarely reported. Another note provided a rare record of house martins and sand martins nesting in holes in a wall, in this case the wall of a sandstone barn on a farm in Devon, UK, without any mud construction on the part of the house martins or burrowing by the sand martins, again showing how opportunistic these birds can be.[3]

For me, one of the most satisfying aspects of observing the behaviour of local birds is experiencing the changes through the seasons: a murmuration of starlings circling over a frosty reedbed on a winter's evening; a flock of sand martins feeding low over the water in early spring; the first swallow of the year; the piercing screaming of swifts in summer; fieldfares and redwings descending on vegetation around the lake to feast on berries in autumn. As I write, in September, the swallows are gathering on overhead wires, twittering incessantly, before setting off for their winter quarters in South Africa. Arrivals and departures of birds throughout the year are a reminder that our local patch is linked to the wider world and that we share birds with other communities.

First arrivals in spring are always a joy, marking the welcome transition from cold and grey winter days to, hopefully, blue skies and warm sunshine. In some years we have to wait longer than in others, however. I was living in Scotland in 1979 when a particularly cold and snowy spring lasted right into May. My local swallows arrived late and, when they did, they hung out at the loch during the day and huddled in the farmyard barns at night, not getting down to breeding until June. This year, 2021, too, spring migrants including swallows were delayed by cold northerly winds in the UK and Europe. The long-term trend in our warming climate, however, is for migrants to arrive earlier. This is evident in our own local patches, although the pattern is clearer when seen over a large area, regionally or nationally. Annual records of migration from the same area contributed by birders and other members of the public to local bird reports and citizen science projects, such as Nature's Calendar and BirdTrack in the UK, are invaluable indicators of changes in phenology, the timing of natural events. Compared with the 1960s, for example, swallows are arriving about 15 days earlier and laying eggs 12 days earlier.[4] One study based on local bird reports over 56 years (1950–2005) found that sand martins had advanced their arrival date even more than swallows; historically they used to arrive after swallows but now tend to arrive first.[5]

As well as variation in phenology, observations in our local area reveal the ups and downs of bird populations. When we are suffering a cold winter, for instance, species such as wrens are too, and we will hear their exuberant song less often in the following spring. A cold winter and poor crop of fruits and seeds in Fennoscandia, however, may mean more waxwings arriving to eat berries on the trees in my local supermarket car park (though not usually the one I shop at!). In the spring and summer too, our local area reflects conditions elsewhere. Fewer swallows survive, for example, when there is little rain, and hence poor vegetation growth and a dearth of insects, in their winter quarters and on migration: another reminder of how our local birds are linked to environments elsewhere. Over the long term, our local records show how populations are faring and are particularly valuable as part of national projects such as BirdTrack and British Trust for Ornithology (BTO) surveys such as the Breeding Bird Survey (BBS), which track the wider distribution, abundance and nesting success of species. House martins in the UK, for instance, which have declined by more than 50% since the 1960s,[6] were the subject of a BTO citizen science survey in 2015 and a nest study in 2016 and 2017. These identified factors improving the success of nests such as the presence of livestock; in addition, breeding success was better in artificial nests than natural ones.[7] As well as contributing to science and conservation efforts, participating in surveys such as these I find is personally rewarding, providing a sense of purpose to local birding, another factor improving both physical and mental wellbeing.

Citizen science surveys have charted the decline of many bird populations, mostly farmland and woodland species, over recent decades.[8] Cuckoos, in particular, I rarely hear in my local area now. Others, often wetland species, have increased. When I moved from Scotland to the south coast of England 30 years ago, little egrets were just starting their invasion of Britain from continental Europe, arriving in autumn to overwinter but rarely staying longer. I often walked at Cuckmere Haven near Brighton, where the occasional white bird with bright yellow feet wading at the edge of the river stood out incongruously. Then they were confirmed breeding in Dorset in 1996 and their numbers took off. I live further north now, in the Midlands, yet can expect to see them on any wetland walk, winter or summer. Now it is likely to be a great white egret or the odd cattle egret or spoonbill that diverts my attention from more common species. This changing mix of the familiar and the new is always interesting and one of the reasons I keep walking in my local area.

On my walks I notice how changes in the climate and landscape are affecting the behaviour of birds. During recent hot, dry spells house martins and swallows have been unable to collect mud for their nests, while they have found it difficult to feed in the increasingly heavy rain. On farms, they benefit from the insects found in new pollinator and wildflower strips left around arable fields, but, as grazed pasture is a good source of flying insects, they suffer where livestock are housed indoors. Our efforts to combat climate change also have an impact on habitats. Crops grown for biomass and solar farms are increasingly common in my local area and a wind turbine was built in the middle of my BBS square a few years ago! How will birds cope with changes in the weather, in their habitats or in biodiversity? Can they adapt or will they disappear? What new species will colonise? To answer such questions, we need, more than ever, to understand the behaviour of our local birds.

5

Long-term Local Science

Ben Sheldon

Like many birdwatchers who have ended up pursuing research on birds as a career, I went through a phase of considering tropical birding as the pinnacle of my ambitions for both birdwatching and research. I was a student at the University of Cambridge in the late 1980s and there was a strong tradition then, among the student birding crowd, of planning 'expeditions' to survey relatively poorly known parts of the world during the long summer vacations. I was lucky enough to join teams that surveyed Marojejy in Madagascar in 1988 and Barito Ulu in Central Kalimantan in Borneo in 1989. They were fantastic experiences: the first time I had been outside Europe, and the first time I had been immersed in studying very new avifaunas with high degrees of endemism. On top of this, the remarkable diversity of mammals (we recorded over 50 species of mammal at our forest site in Borneo), reptiles, amphibians, insects and plants was startling.

The optics of these expeditions are perhaps not so uncomplicated today as they seemed then; looking back, I can remember rather little awareness, at least for myself, that this approach might have overtones of colonialism, or that we were pretty fortunate to be able to contemplate such trips at all, from the perspective of student maintenance grants and no fees. Having said that, we did make efforts to work with conservation and government organisations, and ornithologists, in the countries we visited, and the expeditions involved great commitment before, during, and after the trips. In each case, a small group of students in their late teens or early twenties raised several thousand pounds in funding, planned survey work in detail, travelled on a shoe-string budget, spent two to three months undertaking very intensive fieldwork, and then wrote the results up in extensive reports and papers in the ornithological literature, while simultaneously pursuing their academic studies.

Before my student days I was pretty sure that I wanted to become a scientist; I just was not sure in what field. At school, our biology curriculum covered very little ecology, evolution or behaviour, and cell biology and genetics seemed to be where most of the excitement was. I had read *Curious Naturalists*—a wonderful book about the early days of ethology by Niko Tinbergen—one summer holiday, but it did not seem to square with my school lessons, and it was not until lectures at Cambridge from Nick Davies, and another summer spent reading *An Introduction to Behavioural Ecology*, that I really understood that the sort of field biology that I had always loved could lead to an active area of research. The expedition experiences were enough to cement the ambition in place and I started looking for PhD positions.

The process for finding a PhD was much less formal and organised than now (no internet; just the odd paper advert stuck to notice boards and very much reliant on recommendations and suggestions from academic staff or graduate students one had got to know). I somehow stumbled into being offered a PhD place with Tim Birkhead at the University of Sheffield (no formal interview—just a trip up on the train for 'a chat'— with a letter offering me a place arriving a few days later).

Tim was, at the time, immersing himself fully in trying to understand the evolutionary causes and consequences of sperm competition in birds, a field for which new ideas and technology—notably DNA fingerprinting to assign parentage—were revolutionising the field. Tim suggested that I might study the behaviour and mating outcomes of a common British bird—the choice of species was up to me, but I remember blue tit, blackbird and chaffinch being suggested.

This felt like a fork-in-the-road moment. I had had wonderful once-in-a-lifetime experiences in Madagascar and Borneo (and some of the memories are still sharp now more than 30 years later). To the extent that I had thought about it, I had envisaged distant landscapes and hyper-biodiverse areas being where I would carry out research. I had even dabbled in research a little on these trips, particularly in Madagascar, where I had collected a large data set on bird mixed-species flock composition, and spent several days following and trying to observe behaviour in diademed sifaka troops. Several of my colleagues on the expeditions were already pursuing research in the tropics, or subsequently became influential leaders in conservation linked to the sites we had visited. Here, instead, I was having the prospect of studying the behaviour of a common British bird, in a suburban park on the edge of Sheffield, dangled in front of me.

One thing that helped to tip the balance was another experience while I was an undergraduate. For many years, in early January, the Edward Grey Institute in Oxford ran a student ornithology conference, at which students (and some more senior speakers) gave talks about research on a wide range of topics, presided over by the genial figure of Chris Perrins, and united by the common theme of field ornithology. Attending from my first year at university onwards, I made many friends and connections that have lasted through the years and for the first time experienced the intense stimulation of debating unresolved scientific questions over drinks in the pub. The talks that stood out were as likely to be about common and familiar birds in the UK—dunnocks, great tits, corn buntings, lesser black-backed gulls—as those from further afield. It was clear that you could tackle interesting scientific questions with fieldwork on common species.

I chose to study sperm competition (the behaviours associated with mating competition, and their outcomes) in chaffinches at Whirlow Park, on the western edge of Sheffield. I think I had rather looked down my nose at the prospect of research involving blue tits (not a 'proper' field study species, as it involved nestboxes), and more work had been done on blackbirds; save for comprehensive ethological work by Robert Hinde in the 1950s, rather little work on the behaviour of chaffinches had been carried out (at least, published in English). Thus began a three-year period of establishing a colour-ringed population, and intense observations of individual birds from February through to July, with analysis of behaviour and DNA profiling of samples during the rest of the year. I learned a great deal about how to do science, but also about basic fieldcraft— chaffinch nests could be both remarkably cryptic and quite inaccessible, but the high rate of nest predation rather spoiled my hopes of linking my intense behavioural observations to outcomes in terms of paternity. Blue tits might have been a better choice, though Bart Kempenaers in Antwerp, who became a good friend, was doing some wonderful work on that species to which I doubt I could have added much.

An unanticipated benefit of my chaffinch study site was that it became a default local birding patch, and the many hundreds of hours spent there each year inevitably gave me many opportunities to learn about the local birds in great detail. Rarities were few, though they did once include a brief and shocking April alpine swift, but the site harboured a good collection of breeding warblers and finches, and I enjoyed finding their nests and ringing the nestlings each year. With pied flycatcher, redstart and wood warbler in the Limb Valley just behind it, and golden plover, merlin and ring ouzel on the moors of the Peak District, just beyond, the local breeding birds were diverse. Some late spring and early summer days, after finishing observations and data collection in the early hours, would be extended with long walks out to Stanage Edge, returning via Redmires Reservoirs back to Sheffield along the Porter Valley. I found the long walks over the open moorland landscape a great help in turning over ideas, or research puzzles, in my head.

From Sheffield, I was lucky enough to be offered a postdoc for a couple of years, based at Uppsala University in Sweden, working on life history and disease interactions in collared flycatchers. This was a nestbox study population, established by Lars Gustafsson on the Baltic island of Gotland in 1980, and by the early 1990s it was becoming well known as a really exceptional population for following the breeding success of individual birds from birth onwards. I had shed my doubts about the merits of nestbox studies by then and, as an exciting vagrant to the UK, the collared flycatcher had an undeniable draw (the population on Gotland is an odd northern off-shoot from the main range of this species). This was the start of a period of great scientific productivity for me: Uppsala in the 1990s was a global centre for behavioural ecology, under the benevolent leadership of Staffan Ulfstrand, and I found it extraordinarily stimulating. It helped that the fieldwork was so enjoyable as well. Each year, a small group of us would catch the ferry over to Gotland in late April and spend the next ten weeks in small out of season holiday cottages near the village of Burgsvik. Initially the work involved just regular checks of many hundreds of nestboxes erected in over a dozen deciduous woodland patches. The sight of the first stunning black-and-white, male collared flycatcher, dancing butterfly-like among the still mostly leafless trees, against a blue sky and over a carpet of wood anemones was a spectacular annual treat, but at quieter times for research we could enjoy great local birding with hordes of migrating geese, ducks, waders and terns on the nearby marshes, and sometimes spectacular falls of migrant passerines on coastal headlands, with a liberal dose of rarities (a bimaculated lark one May afternoon was a real highlight).

The scientific merits of a long-term population study, like the Gotland collared flycatcher study, are extensive, but the most important are as follows. First, the standardisation of data collection enables comparison of the same variables over time and hence understanding of change over time; consequently the data has been invaluable for understanding the effects of climate change and other human-driven changes in nature. Second, the long-term nature of the study in itself means that a really representative sample of the population is obtained: one can understand what the effects of unusually high or low densities, food abundance or scarcity, or climate extremes are. Third, identifying individuals means that accurate data on variation in fitness—the raw material on which evolution depends—can be obtained. In birds, an added advantage is that it is more straightforward than in many organisms to link together parents and their offspring by simply catching parents at the nest and also ringing their young. The Gotland study was scientifically successful in part because both adult and nestling flycatchers showed high degrees of philopatry— the tendency to return to the same place (site of previous breeding for adults, and of birth for nestlings). It is something of a puzzle why this tendency is so high in this population, but lower for the closely related pied flycatcher which is otherwise so abundant elsewhere in Fennoscandia. Perhaps it is linked to the isolated nature of this population—which is an 'island' of collared flycatchers in a sea of pied flycatchers—and for which philopatry may be strongly selected. It was certainly notable that we saw collared flycatchers quite rarely in the big falls of migrants on the Gotland coast, even though they were among the most abundant breeding species in woodlands only a few kilometres away; in contrast, pied flycatchers were frequently seen migrants, but bred at low densities in the woods on Gotland.

My experiences on Gotland really brought home to me the value and potential of science based on intensive long-term study of common species, and when it was time to return to the UK, on a Royal Society Research Fellowship, I was naturally drawn to Oxford. In 1942, the ffennel family had gifted the Wytham Estate, just outside Oxford, to the university to be used for research and 'the instruction of suitable students'. Comprising over 1,000 hectares, centred around a calcareous hill, its best-known feature is the 385-hectare mixed deciduous woodland that cloaks the slopes of the hill, running down to the Thames which loops around its northern edge. It was here, in 1947, that David Lack established what is, if not the earliest, surely one of the best-known long-term studies of birds in the world, with his population study of the great tit. Lack had recently been appointed to a post heading the Edward Grey Institute, from Dartington Hall in Devon, where, as a schoolmaster he had carried out his famous studies of colour-ringed robins. Seeking a species that he could use to understand the basic demographic nuts and bolts of a population—birth and reproductive rates, lifespan, dispersal—he used the insights that Dutch ecologists, particularly Hans Kluyver, had garnered from studying great tits in the Netherlands, to establish his own population study of great tits in nestboxes at Wytham. Initially in just part of the woods, it was extended to cover the whole site from 1958 to 1961 by Chris Perrins, and has remained effectively standardised ever since. With decades of great tit life histories, as well as a large population of blue tits, it was a very attractive site in which to put down some scientific roots.

Wytham has become a very important place to me in the two decades I have worked there. Overlooking Oxford, it offers a stark contrast with views of the bustling city and the ever-busy A34 from the top of the hill, but can be reached in 20 minutes by bike from the centre of the city. Increasingly popular with local people since the first Covid-19 lockdown of 2020 (anyone can apply for a walking permit to gain access to the woods), it is large enough that there are always quiet rides and patches of woodland, far from most people, particularly along the edge of the Thames. It is biologically not extraordinary: it is a good representative of mixed deciduous woodland in lowland southern England, bearing the marks of centuries of human activity. It has some stands of ancient woodland, but also many patches that are eighteenth- or nineteenth-century plantations or secondary regeneration, and we await the devastation of ash dieback which is expected to lead to the death of the majority of individuals of the numerically commonest tree. Wytham has a high diversity of insects, particularly Lepidoptera, but that diversity is largely as expected given its geographic location and the intensive study of the site. Over the past six decades there has been plenty of turnover in its breeding bird populations: woodcock, redstart, nightingale, willow tit and lesser spotted woodpecker have all vanished—in fact they are pretty much extinct in this part of England. Buzzard, red kite, raven and firecrest have all (re) colonised; marsh tit maintain a healthy population, perhaps benefitting from food provided in winter by researchers aiming to catch commoner tits to mark them. Numbers of the main study species, great and blue tits, have fluctuated over the last six decades from just over 100 breeding pairs to close to 500 pairs of each species, but in an average year there are around 700 breeding pairs of tits in just over 1,200 nestboxes.

What makes Wytham special is the accumulated body of knowledge both in depth and breadth across the site. Each year, we add new observations of great (and blue) tit lives, loves and families to an expanding data set; each year adds a layer of new data. Despite approaching 75 years of data, each year is subtly different, but together they build up a comprehensive picture of how a population has changed and responded to the many changes in the environment. But, even more exciting, any single great tit might be connected through its relatives to dozens of others alive in the woods at the same time, and its ancestry can be traced back, through many generations, to birds that were alive many decades ago (the longest lineage spans 38 generations). In the winter, the birds form loose flocks; highly dynamic in composition, study of their make-up shows that not only are birds connected to many others in terms of kinship, but also by social relationships. These two networks create pathways along which information flows: we try to understand how genes and the environment, on the one hand, control the variation we see among individuals. And we also seek to understand, on much shorter timescales, how social networks lead to information about food, predators, and social opportunities spreading among individuals and changing behaviour.

Allied with this rich, multi-layered data set on great tits, is a wealth of information about other aspects of the ecology of Wytham. From soil chemistry (a 10,000-fold calcium gradient over little more than a kilometre has implications for a bird that lays more than its body weight in eggs each year), through invertebrate population dynamics, to the diversity, structure and health of the forest trees, we increasingly find ourselves connecting together the bird data sets with those from other systems. Doing so helps us to try to understand how birds respond to signals and their environment on many scales, especially when these might be acting in opposition. And the accumulated body of work on the great tits of Wytham would fill many books—over 400 scientific papers and more than 40 PhD theses have been based, at least to some extent, on these birds.

Returning to myself as a 20-year-old student birder in Madagascar, could I have predicted that my focus in research for the last 20 years would have been on something so commonplace and every day? I do not think I had appreciated at the time that, while seeing novelty in terms of diversity of bird species would require ever-more travel to more and more places, this is just one dimension of diversity. There is diversity in behaviour, in individual lives, in parasites, songs, learning, social relationships and genes, all of which occur in great diversity among those species that we find around us. And making sense of this diversity needs the kind of multi-layered data that come from adding the study of birds to other aspects of ecology of a site. That the potential for this kind of study is literally on our doorsteps is a truth that has taken me time to appreciate, but that I hope to go on appreciating for many more years.

6

The Perpetual Patch

Roger Emmens

It is something that thrills you anew each time you experience it: the glint of the rising sun on water, the rustle of reeds, the mechanical croak of gadwall, the scratchy rhythm of reed warbler, even the clamour of the black-headed gull colony.

Many birders know and treasure their local patch, familiar with all its birds and yet always hopeful of something out of the ordinary. Because it is local, you can pop down there whenever you have the odd spare hour or two, and so you get to know all its moods, know what birds to expect, and more easily spot when there is something out of the ordinary. So it is with ringers too.

Rye Meads Ringing Group (RMRG) is one of the longest established groups, being in our 62nd year of operation at Rye Meads in Hertfordshire. The group was founded in 1960 to ring and study the birds on the then newly established sewage treatment works, which had a large series of settlement lagoons nestled on the flood plain of the River Lea. These founder members were delighted to have such an interesting area on their doorstep, and so are the current active members—it is now *our* local patch.

From new lagoons with bare banks to today's reedbeds, scrub and wet meadows—half now leased to the Royal Society for the Protection of Birds (RSPB) for its Rye Meads reserve, with an area of Herts and Middlesex Wildlife Trust (HMWT) reserve—as well as the still very much operational sewage works, the site has changed hugely over those 60 years. And we have been there to record it all.

This is not just about ringing birds. Of course that is an enjoyable and worthwhile pastime in itself, but it is only the beginning of the process. The particular value of a group like ours is the focus on a specific site, showing the benefit of combining ringing, nest recording, breeding surveys and general observations over the long term for demonstrating changes in the site and in its importance to different groups of birds, as well as wider population shifts.

When I started with RMRG in the 1980s, we were regularly ringing our local breeding species such as turtle dove, tree sparrow, willow tit, willow warbler and lesser redpoll. We could catch winter flocks of corn bunting and linnet, and spring passage of greenfinch would yield a few hundred captures. Sadly, all no more: it has been 19 years since we even recorded a passing willow tit.

But, on the other hand, we would rarely see any raptors other than the occasional kestrel, certainly not the seven species now regularly seen. We never had any little egret or great white egret or raven; Cetti's warbler was an exotic rarity and gadwall was an uncommon visitor.

That is the point about a local patch. It changes. We know every inch of our 97 or so hectares. The habitat here needs constant maintenance—another job the ringing group does its share of, together with the RSPB and HMWT volunteers and staff—and conditions vary from year to year, so it is not just the national population fluctuations that affect us. Observing and recording these changes is part of the fascination, and when you have more than 60 years of ringing data and observations, there are plenty of ways to analyse the changes.

Perhaps one of the most important activities RMRG undertakes is our annual breeding bird survey. We cover the whole site, mapping singing males and counting territories, to ensure that we can see year by year how our local populations are changing. We have got a pretty good idea of how our local birds are faring, and can compare this with the national data reported in British Trust for Ornithology (BTO) surveys, so we know if any changes locally are taking effect. In the process we conduct much of our nest finding and recording, and that enables the ringing of chicks, which is especially valuable. And it also dovetails with our Constant Effort Site ringing to help measure breeding success.

And what do you do with all that data? Well, you do not just assume that some scientist at the BTO headquarters will use it for some sort of national study: you try to use it yourself to see what is going on with your own birds. We produce a report every three years, and over the lifetime of our group we have published in our reports some 80 papers on widely varying topics, from greenfinch weight variation, to pullus (ringed in the nest) sedge warbler departure dates, to wintering chiffchaff patterns, to changes in wing lengths in willow warblers, to the status of our wintering water pipits.

But we do not just look at our birdlife. Through the varied interests of our members over the years, we have studied the wildflowers, butterflies, moths, bugs, molluscs and bats, among others, of Rye Meads. You do not have to be a chiropophile to be interested to learn that there are ten species of bat regularly using Rye Meads airspace.

That is the point of being part of a patch ringing group. You develop a proprietorial interest in all of the natural history of your patch. You want to know how it all works together. For example, casual observation revealed that the clumps of fat hen that spring up on disturbed ground on the water margins are favoured food for some ducks, so we try to keep these clumps going instead of 'tidying up' our water margin net sites.

Why do linnets no longer breed on site when they still breed nearby? Is it the habitat? If so what about it has changed? Are there certain insects or seeds that they require that for some reason are no longer found here? Or, why did black-necked grebes breed here successfully for one year and not thereafter? Are we perhaps not managing water levels in the right way, or is it something else? And another thing: why do some years give us large swallow roosts in our autumn reedbeds, and other years yield just penny numbers? It does not seem to be related to national breeding successes; but if it is weather conditions, what exactly is the determining factor?

There are always questions when you study your patch, and one of the pleasurable activities of the group is to sit over a mug of tea after the fieldwork is finished and chew the fat about unanswered questions like these, to see whether anyone can come up with a theory and a way to test it.

And this shows the value of group effort and regular counting and surveying over a long period. Whenever a potential investigation comes to mind, we already have the back data. Take the decline of the house sparrow—one of the most poorly documented declines of a common bird. We know exactly when it started to decline at Rye Meads (slowly through the 1970s, main decline in the early 1980s), and when it ceased to breed here (1998). We used to get winter flocks in the hundreds; now we get more records of marsh harrier or raven a year than house sparrow.

Because we log our observations every time we come to Rye Meads, we can see exactly how and when our birds change. We do not need to see warning signs to begin collecting data: we already have it. We constantly have cause to thank our predecessors in the group for all the data they collected before we came along and took up the baton!

We saw the decline of seed-eaters happening through the 1980s, when our populations of corn bunting, lesser redpoll, bullfinch and tree sparrow fell sharply. We still get a few lesser redpoll on passage, and a couple of pairs of bullfinch still breed, but corn bunting and tree sparrow are extinct here. Similarly we recorded in detail the expansion of willow tit from occasional visitor in the 1960s to breeding regular in the 1970s and early 1980s, and then sadly its decline and loss even as a rare visitor. We watched the arrival of Cetti's warbler in 1975, first breeding in 1978, and bemoaned their being wiped out by the severe winter of 1980–1; then we welcomed their re-colonisation some 26 years later!

As one of the few locations ringing many ducks, and especially ducklings—which provide especially valuable data, being birds of exactly known age and origin—we have made discoveries about how our tufted ducks migrate. We have ringed getting on for a thousand, including more than 500 ducklings, and can see a pattern in reported recoveries (sadly mostly birds shot by hunters). It is a shock and a delight to discover that a duckling ringed at Rye Meads can turn up as a breeding adult more than 4,000 kilometres away in Russia! It seems that tufted ducks pair up on their wintering grounds, and the male follows the female back to her natal area; foreign males may stay with our local females to breed here. So, birds from areas that are frozen in winter, from Iceland to Siberia, come to Rye Meads for the winter, and some of our locally bred males travel the other way with their new wives! Incidentally, this mechanism for genetic mixing demonstrates why ducks tend not to develop regional races or subspecies.

As chiffchaffs started to spend winters in Britain more and more often, we were well placed to monitor this trend. Because we work on chiffchaffs all year round, we were able to prove that our wintering birds are not the same birds present in other seasons. We had the vital data, we just did not know it was vital while we were collecting it, until it proved useful to answer one of those nagging questions debated over that mug of tea. Furthermore, our recovery data, while not conclusive, indicates that our passage and summering birds winter in southern Europe and north Africa, whereas it suggests our wintering birds originate in continental Europe, east and north-east of Rye Meads, from Belgium to Denmark.

Another example of already having the data and using it to explore new unanswered questions is the decline of the willow warbler. A well-documented collapse in populations in southern Britain occurred during the 1980s for reasons that were not clear. Our breeding numbers also dropped dramatically, although we still ringed reasonable numbers on passage. A bird such as the willow warbler migrates to sub-Saharan Africa, undergoing a very hazardous long-distance journey, and so it is to be expected that the normal physique of such a bird reflects a compromise in the very different demands of feeding, predator evasion, breeding and migration. When one of these factors becomes distorted, then the compromise may be disturbed so that if, for example, longer-winged birds (presumably, therefore, larger birds) found greater difficulty in finding enough food in the winter quarters, then the population as a whole might become shorter winged, even if this put the individual at a disadvantage compared with longer-winged birds for the other demands of breeding, migration and predator evasion.

We used data from the ten year period 1965–74, which was before the population decline, and 1987–2006, which came after. We found evidence of a statistically significant reduction in average wing length of returning spring adults which cannot be explained by racial or sexual differences, and so might represent a change in survivability while on migration and in winter quarters which favoured shorter-winged individuals.

That, naturally, then begs a further set of questions: are these changes temporary or are they sustainable? Are there morphological changes in birds from Scotland, which has seen a simultaneous rise in population? Are the birds passing through Rye Meads in spring destined to breed locally or further north? We need more data!

Which, of course, means that it is as relevant as ever to keep monitoring the birds on our patch and keep collecting data on them. The increasingly uncertain future, with global warming, pesticide impacts and pollution all increasing pressures on bird populations, means that the local patch is as worthwhile as ever as a destination.

7

The Long Rhythms of a Place

José Ignacio Dies Jambrino

The marbled teal emerges discreetly from the vegetation and after a few seconds disappears, not to be seen again that morning. Five flamingos rest in the middle of the lagoon; in the distance, the traffic of terns from the colony to the sea and the paddy fields continues uninterrupted until suddenly it explodes into a frenzy of alarm calls and defensive flights. It is late June already. Could it be an early Eleonora's falcon? A few seconds later the tern traffic is restored and life returns to normal. The flamingos had not even been bothered, but the black-winged stilts keep looking up nervously for the next few minutes.

I have spent three decades witnessing this familiar scene and yet I still feel a sense of disbelief. In the late 1980s I could have only dreamed of witnessing some of these birds breeding or regularly present in my own patch in l'Albufera de Valencia, one of the largest coastal lagoons in the Mediterranean; and I could not have imagined the unexpected turns and twists of a journey which is as much about the restoration of a wetland as it is about becoming a better birdwatcher.

A pivotal moment in this journey was the spring of 1990. I was still a student when I was asked to accompany a visiting British consultant who was collaborating with the local government to draw up a preliminary project for the environmental regeneration of the Racó de l'Olla reserve. His name was Herbert Axell, something of a legend, I had heard, because of his reputation earned at the Royal Society for the Protection of Birds' Minsmere nature reserve.

I remember the point in our conversation when Mr Axell assured me that the avocet would certainly breed in the reserve. I looked at him in disbelief—not only were there no historical breeding records but the bird was a scarce migrant. He also told me about the importance of extending the saltmarsh, of not fragmenting it, of understanding the geomorphological essence of the reserve, the characteristics of the soil, the water, the landscape. He talked about the need to accept the inevitable ecological succession, as a process of continuous change in the structure of the ecosystems that we can harness to our advantage. At the time, the contrast between what lay before my eyes and the future landscape painted by Mr Axell could not have been greater. The area had been used as a salt pan until the seventeenth century, was later regarded as a wasteland as it was not suitable for growing rice, and in the 1960s had been partly filled with rubble in an unsuccessful attempt to build a racecourse for horses.

The actual regeneration plan was the responsibility of a team of local ecologists. Their original plan led to the re-creation of diverse lagoon environments to enable the breeding of locally endangered waterfowl, and small islands for breeding colonies of waders, gulls and terns. The work was completed in January 1993 and it was then that I began serving on the management team of this area. After almost three decades of witnessing how that daring attempt at environmental restoration has unfolded, I am still surprised at how little of what has been achieved actually resembles the original plans.

The original project sought to enhance the biological richness of the reserve by adapting various freshwater and saltwater environments with different levels of flooding regulated by an electric pumping system, a process that was both labour intensive and expensive. This approach, however, was eventually abandoned not long after the regeneration works in favour of a more dynamic management style that considered the whole reserve as a single, coherent brackish environment, whose diversity would be defined by the variable extension of flooded or exposed surfaces throughout the year, allowing natural summer drying, even if it was total, and winter flooding according to rainfall. This decision proved to be the right one and over the years this management style, based on the principle of minimum intervention, has led to the establishment of a resilient Mediterranean saltmarsh environment.

In terms of the diversity of species and number of breeding birds, the results have far exceeded our initial hopes and expectations. Confirming Axell's prediction, eight pairs of avocets raised chicks during the first breeding season after the regeneration works, and the breeding population reached a maximum of 126 pairs in 2001. In addition to this, a total of 27 species of waterfowl have successfully bred in the reserve, eleven of these species being new to l'Albufera. Species that were locally scarce, such as the common tern, have fared well, and we have recorded a total of 51 breeding attempts by the endangered marbled teal, which maintains in the reserve its northernmost regular breeding locality in Europe. Overall, the reserve has hosted 118,000 water bird nests from its establishment until the present day. An annual breeding population that has averaged around 4,000 pairs, makes it the second most important breeding locality in the Iberian Peninsula for some species of gulls and terns, after the Ebro delta.

Behind these numbers there is an ongoing process of change in the composition of the population of breeding birds. Part of this change has been a response to changes in the habitat. Little terns, black-winged stilts, Kentish plovers and collared pratincoles were pioneers in colonising the reserve. A few years later, the colonies of common terns and black-headed gulls numbered several thousand pairs. From the early 2010s, the combined population of species that had not bred here before—gull-billed terns, sandwich terns, slender-billed gulls and Mediterranean gulls among others—exceeded a thousand pairs.

This colonisation sequence shows how these species respond to changing environments. Little terns and plovers take advantage of new, unstable habitats such as new sandbanks, still devoid of vegetation, with little guarantee that a blow from the sea will not sweep away their nests. Larger terns and gulls use these same spaces later, once they have become consolidated.

Changes in the community of breeding birds have also occurred as species that once successfully occupied the reserve were displaced by larger species, in a sort of conservation paradox. Avocets began to displace black-winged stilts as both species vied for the same habitats, black-headed and slender-billed gulls stole the fish brought by the terns for their young, and a few gull-billed terns became selective predators of plover, tern and pratincole chicks. And as larger species moved in they claimed larger parts of the habitat. The greater flamingo was initially absent but the number of birds increased from around 100 in the early 2000s to around 1,000 in 2012 and around 5,000 birds in 2017. Flamingos use the reserve for resting and feeding, competing for the same resources on which the young of other species, such as marbled teal, shelducks or avocets, depend.

What happened at the Racó de l'Olla reserve provides an insight into what might have been a normal pattern of colonisation and displacement of breeding birds when coastal areas in the Spanish Mediterranean were a long succession of shallow lagoons separated from the sea by newly formed and shifting sand banks. In fact, it was while trying to make sense of these dynamics that we learned that centuries ago, before becoming a salt pan, what is now the reserve had been a sandbank in direct contact with the sea in one of the inlet channels joining the coastal lagoon to the Mediterranean.

As I write this account of simultaneous change in the landscape and the population of breeding birds I am fully aware that the process sounds all too smooth. Retrospectively it may even look inevitable. The reality, though, is that it was an uncertain journey punctuated by many questions about what steps to take next, what species would benefit and what species would lose out as a result; minimum intervention is, after all, still intervention. We proceeded by tinkering, trial and error, constantly reminding ourselves that there may be dynamics that we were not aware of or would not fully understand. We were restoring a place but we were not fully sure what parts of its former pasts we were restoring. Humans had been working this land for centuries and layer after layer of soil bore evidence of different land uses. By observing how the place responded to the major restoration works of the early 1990s and, after this, the small-scale interventions, we noted that some ecological and geological features took precedence over others, and we adapted our understanding and plans for the reserve accordingly.

Patient observation of a particular site and the opportunity to tinker with it have enabled me to understand birds in context: each record, each observation of behaviour adds to a wider, growing but always partial picture of the place and its wildlife. There is knowledge about a place, its life and its rhythms that you acquire with time, knowledge that becomes a part of you. It is simultaneously an ability to read the subtleties of a landscape—a disposition shaped by thousands of hours and days patiently observing its life unfolding—and the humble realisation that there is still so much that you do not know; that no matter how much data I could collect about this place I would still be grasping only some of its dynamics. This knowledge and this idea are difficult to put into words but it is something that seasoned patch birders will understand.

That somewhere like the Racó de l'Olla reserve can exist a mere 16 kilometres from the third-largest city in Spain is yet more proof that good sites for birding can be created virtually everywhere. It is also proof that the large-scale regeneration of habitats that are needed to reverse the decline of biodiversity can benefit greatly from the skills and knowledge developed by patch birders.

8

A Life of Local Birding

Matt Phelps

Birding, for me, very much began at home. I did not really know what birding was in the beginning, let alone low-carbon birding. I just knew that I enjoyed watching the various birds that visited the family garden. My childhood and teenage notebooks reveal somewhat crude descriptions and sketches of common species such as dunnock and great spotted woodpecker, but I still took pleasure in identifying them and observing their behaviour. It was my dad who first drew my attention to the swifts whizzing and screaming over the house, and I can recall many summer evenings spent in a deckchair learning to pick out the house martins among them as they swirled around, impossibly high and seemingly effortless in their mastery of flight.

As my interest developed, walks and cycle rides from home allowed new opportunities for exciting wildlife encounters. The nearby nature reserve where I played as a child became a whole new source of education as I began conservation volunteering and learning about the array of exciting birdlife there. One particularly memorable event involved being taken to see a sparrowhawk nest in a quiet corner of the woods, the heads of the chicks clearly visible in the spring sunshine as blackcaps and chiffchaffs sang all around. It was here I also saw my first kingfisher, red kite and siskin, among others.

Growing up on the edge of the Thames Basin Heaths Special Protection Area meant visits to nearby heathland sites, where seemingly alien species such as nightjar, Dartford warbler, cuckoo and woodcock revealed themselves. I remember one particular evening cycling across the military ranges just a couple of miles from home and having the incredible experience of a nightjar buzzing right over my head as I pedalled through the darkness. Sometimes I would ride along the towpath by the Basingstoke Canal where I would pass fishing common terns and grey herons, or visit Farnborough airfield where I saw and heard skylarks and tree pipits parachuting above my head.

My love for birds wavered a little in my late teens and early twenties, as so often seems to happen, but when career choices took me to agricultural college it reawakened my interest and I returned to the local heaths for more nocturnal avian adventures and later began visiting my first proper local patch, Tice's Meadow, on the Hampshire/Surrey border. Here I discovered even more surprising species to be found just a couple of miles from home, including many different ducks and warblers, lapwings and rarer waders. I was well and truly hooked.

Early in my career I was employed for several years as a gardener at a large garden near Guildford and it was here, working mostly alone, that I really began to connect with birds on an intimate level. I noticed every arrival and departure, every nest and every subtle change in the garden. I learned that the 'different' pigeons that flushed from a hole in an old horse chestnut tree every time I passed were stock doves, I learned to recognise the difference in drumming between great and lesser spotted woodpeckers and delighted in every visit to the pond by a kingfisher. The more I learned, the more

I wanted to know, and every day felt filled with possibility as I searched the corners of the garden for new and exciting birds and other wildlife. This was the first time I really became aware of the chance of stumbling across rare and scarce species in a familiar setting, as more unusual visitors included waxwing, spotted flycatcher and sedge warbler. I also began the habit of routinely gazing upwards, realising the sky could hold just as many surprises as the land, and was rewarded with flyovers from species such as osprey, golden plover, Mediterranean gull and even a gannet! The wonders of visible migration were a whole new concept to me at that time, but it is something I have been fascinated by ever since.

In 2014, I joined a team of local birders led by the ranger at Leith Hill, Sam Bayley, to carry out regular visible migration studies from the top of the tower on the hill, the highest point in south-east England, no less. This has been something I have carried on intermittently to this day and again has proved the value of spending time in one spot (quite literally in this case as the top of the tower is only around five by five metres) as among the more unlikely birds to have passed over this part of the Surrey Hills are brent goose, great skua, little gull, osprey, marsh harrier and great grey shrike. It is not always about the rare species though, as sometimes the most memorable moments can come from the more familiar, such as the time a wheatear very nearly landed on my head atop the tower!

The intimate level of connection with a place and a landscape is one of the key things that has kept me mostly local in my birding habits since those early days of watching the garden birds as a boy. I enjoy the occasional trip to another county or the coast, but the excitement of heading out locally to find my own birds is something that has only gotten stronger.

For the uninitiated, a patch is an area chosen by a birder as their regular haunt. There are no hard and fast rules, but it generally tends to be somewhere close to where the birder lives and of a suitable size that most of it can be covered in a single visit. It can be a designated nature reserve, a park, an area of farmland, a reservoir or even just a garden. For me, patch birding is about building a relationship with a landscape and studying the subtle changes and phenological shifts throughout each year, be it the first burst of willow warbler song in spring or the trickle of swallows and martins that becomes a torrent, or the gentle beginnings of southward movement in the autumn. Avian migration in all its forms holds a special fascination for me, and watching a local patch is, in my experience, the best way to notice those day-to-day developments first-hand. Seeing a wheatear or yellow wagtail, fresh in from Africa, drop out of the sky in front of you is even more special when it feels like the birds have come to you. Despite all the rarities I have seen since, finding little ringed plovers on a duck pond in Woking and a green sandpiper in a flooded farm gateway near the River Wey, moments from home, remain among my most vivid and treasured birding memories.

I have read that some indigenous people can navigate their way home, even in darkness, by following the calls and songs of individual birds, so familiar they have become with their native landscape. While I would not lay so lofty a claim on my own birding abilities, I can well understand how a person can develop such a profound understanding of their surroundings. An acute awareness of the familiar and routine birds in a place means that anything new immediately jumps out at you. It certainly is possible to learn to recognise an individual song thrush by its penchant for including other bird species' calls in its own song. For example, I have only recently noticed that several of the blackbirds near me incorporate a curlew-like phrase in their song, owing to having a wet grassland on our doorstep.

What also becomes more apparent when one concentrates so intently on a particular area, and the comings and goings of the birds it plays host to, are the changes in species diversity and abundance, breeding success and arrival, and departure dates of migrant species, year on year. We know from various studies carried out by the British Trust for Ornithology, BirdLife International and others, that migratory species are arriving on our shores earlier and earlier and, very often, leaving earlier too. In 2021 I heard my earliest ever nightingale and cuckoo (4 and 9 April, respectively), and in 2020 I saw my earliest ever willow warbler (22 March). For now, this gradual advancement of both arrival and departure means the breeding success of migratory species does not appear to be in any way curtailed. Of course, with droughts and extreme weather events likely to become more common as global temperatures increase, particularly in the southern hemisphere where many of our summer migrants spend the winter months, these species will likely find themselves facing ever more challenging journeys to reach us. It is hard not to be reminded of this each year as we see our own seasons and weather patterns becoming more erratic and extreme. House martins seemed very late in arriving in 2021, doubtless held up by persistent cold northerly winds; then, when they did finally arrive, they discovered a landscape parched by one of the driest Aprils on record, and largely devoid of mud with which to build their nests.

Species such as herons and egrets are well documented as being early adapters to the warming climate and many, such as cattle egret and spoonbill, are now rapidly colonising the British Isles, where once they were merely occasional visitors. These days my local patch is Pulborough Brooks in West Sussex, and it is indicative of the changing times that spoonbill is an increasingly frequent sight on the reserve, while black-winged stilt is just about an annual occurrence now, when not so long ago it would have been a highly prized rarity. Indeed, the species bred in Sussex just a few years ago and I am sure will do so again before long.

Of course, while some species can adapt, others are not lucky. In the relatively short time I have been birding I have seen first-hand the declines in species such as turtle dove, cuckoo and swift as habitat loss, food shortages and, of course, climate change, take their toll.

Pulborough Brooks is an incredible place and one I consider lucky to be able to call my local patch; but wherever I have lived, I have always gone out in search of birds on my doorstep. Living in the village of Pulborough it was an easy choice to decide where I would be focusing my birding efforts, but it has become a tradition of mine that whenever I have moved house, one of the first tasks has been to study a map and set out the boundaries of what would be my patch. The Covid-19 pandemic has made us all more aware of the value of finding places nearby to watch birds, and the Brooks has certainly been a great sanctuary to me in extraordinary times; and I am sure it will continue to be a special place to me for many years to come.

Ironically, the all too familiar 'stay local' message coined during the pandemic is particularly pertinent with the climate and ecological crisis looming. In retrospect, I have always been largely low carbon in my birding habits but, as my chosen hobby constitutes such a large element of my life, it seems only natural that I would choose to focus on how to do it in as low impact a way as possible, at a time when we all must become more mindful of the choices we make. The scale of the climate crisis is enormous and terrifying. We all make decisions every day that can have significant environmental impacts, whether we are aware of them or not. We can all look at how we live our lives day by day, and what small changes we can make that might make a difference.

When it comes to modes of transport for birding, I do not think it is possible to beat the feeling of freedom offered by travelling on foot. I currently drive a lot for work during the week, so it is liberating at the weekends not having to think about where I have parked the car or if I will need to stop for petrol on the way home. I have also rediscovered the joy of travelling by bicycle to see birds, something I had not really done since those early days I mentioned at the start of this chapter.

As a kind of homage to the wonder and excitement that local birding can bring, I can think of no better example than 8 April 2021, my birthday. I had taken the day off work with a view to spending the morning birding locally, and, after a few hours out, was considering walking home when I received a message from a friend informing me that a neighbour of mine had just seen a northern mockingbird in her garden. This same individual had been residing in a garden in Devon for several weeks until that point, and was only the third record for Britain. Fast forward to later in the day and there was the bird sitting on my ivy-covered fence just ten metres or so from the bedroom window. One thing is for sure, I was very glad I had not decided to take a day trip somewhere else instead!

9

The Joys of Patch Birding

Maria Scullion

Patch birding can at first seem like a strange concept. Against the apparent allure of a hectic quest for a long list of species spotted in many places, there is a beautiful simplicity in visiting the same place week in, week out, getting to know the individual creatures that live there and watching the environment change through the seasons. Developing the patience to tune in to the subtlety of those changes makes you a better birdwatcher and naturalist. It helps you to improve your understanding of behaviour and interactions between the different species and how these change as the seasons progress. With lists that you build each year, you can create comparable data sets which allow you to see trends related to species. Simple examples are recording when you hear the first chiffchaff of spring or spot the first returning swallow swooping high above, and the first skein of pink-footed geese arriving, making a racket, in autumn.

Patch birding is about acknowledging the context. Some birdwatchers with good health and easy access to great places, and public transport or cycling paths, can avoid motorised travel, but for me this is a challenge as I live with two chronic illnesses that can limit my physical activities. Cycling is out of the window for me, which does limit where I can go. I can manage to walk for about eight kilometres on some days, which comfortably gets me around my patch from home, but if I am having a bad day then I have to drive to the wetlands. I have learned the hard way that it is not worth the aftereffects to push myself too far, even if there is something exciting on the patch or someone invites me out at the last minute; I have to understand my limits. Ultimately, in my view, the benefits derived from being in nature outweigh the three miles worth of petrol consumed by my car. The general close proximity of patch birding does make it more accessible, and it is one of the reasons I was drawn to it in the first place.

I first heard the term 'patch birding' during my placement year working at Martin Mere. While living there I had unlimited access day and night to the whole 800-acre reserve, so I guess I can say that the whole place was my patch. I was truly spoilt for choice with the vast array of birds just outside my bedroom window, including local rarities such as black terns, a Temminck's stint and bearded tits to name but a few. The haunting sound of a thousand roosting whooper swans in winter and the cacophony from the black-headed gull colony in summer kept me awake at night. And I grew to recognise the different calls of the birds and what they meant: how whooper swans had a specific *whoop* they would make just before they took off, followed closely by the slapping of their feet on the water.

I had unknowingly started patch birding earlier, while still at university. I had not heard of the term but, due to a lack of transport, I would only watch birds on campus, where I lived. It was a countryside campus with a healthy population of birds such as yellowhammers, wagtails, and the occasional brambling, and it was on this patch that I became interested in bird ringing and started my training with a ringing group. I was fortunate that it was an agricultural campus, so I did not receive too many funny looks while walking around at half past four in the morning on my way to the ringing site carrying a rucksack and a camping chair—something I have later realised gets a lot of unwanted attention when there is no context! Bird ringing is, to me, a brilliant way of learning so much more about birds and understanding their behaviours, feeding habits and so on. I remember the first time I saw a bullfinch in the hand, the pure size of its beak stunned me. But then I realised that the size of the beak makes sense when you know that their diet consists of shoots, buds and fruit. Then if you look at the beak of, say, a goldfinch, it is much smaller and pointier, making it much more specialised for eating seeds and foraging on the ground. Although this is something you can pick up on while birdwatching, seeing a bird in the hand accelerates and deepens the learning process.

Back home in North Yorkshire, my patch includes my house and garden and then a walk along the rivers Skell and Ure, including a wetland reserve. My house and garden are a good place to start. We have house sparrows nesting in the gutters (and on adjoining houses too), a pair of woodpigeons who attempt to nest in our laurel every year (and each year they fail due to lack of nest engineering skills) and starlings that come in great numbers in autumn and winter—we were lucky enough to play host to a murmuration over our house in February 2021. I would estimate there were maybe 20,000 birds, about 1,000 of which decided to roost in our garden a few times, unfortunately causing a significant amount of damage to our laurel tree. Despite this, they did provide a great amount of fertiliser for the garden, some of which was deposited on my neighbours' cars, much to their dismay! At night-time I sometimes stay up late and listen out of my window. You can often hear the familiar call of tawny owls, who frequent the garden and sometimes even perch on the back of our house. Having these wonderful birds in such proximity to the house makes you realise that travelling far really is not necessary in order to see or hear amazing birds. On one February night alone, I heard common scoters and a large flock of whooper swans over the house. Sometimes, even during the daytime, it is worth just taking a moment to listen. It is surprising what you can hear, especially during migration. But listening is also about learning from an absence of sounds. One day you realise that the swifts are no longer screaming in the sky or, in early spring, that the screeching pink-footed geese have stopped drifting over. This changing yet predictable soundscape reminds you that, regardless of what is going on in our lives, the behaviours of birds do not drastically change; their routines signal (at least in my mind) the different points in the calendar.

The rest of my patch is found on a walk along the canal which begins on my street. In the basin you have your standard mallards and the occasional heron, but recently we have had an otter visiting! I think it must be an inquisitive young one as it spends a lot of time following people up and down as they walk, watching from the other side of the water. There is a definite increase in biodiversity along the canal; I have lived nearby for over ten years and it is only in the last year that otters have been spotted on that stretch. I have also noticed an increase in kingfishers this last year, from one pair to two pairs, and a third pair of dippers in the river that joins with the canal. Having regularly walked these routes during the Covid-19 pandemic, I decided to take up the Wetland Bird Survey (WeBS) count for the area, which had been left unrecorded for some years. I enjoyed the structure of it, and seeing new birds each time I surveyed. It was great to see the shelducks and oystercatchers return for the breeding season in the spring and the large groups of geese in wintertime. It was also a good way to improve on my general bird identification skills: there were some species like lapwing that I would see at each survey that I now feel very confident in identifying, both by call and by sight, in the air and on the ground.

It is not just the birdlife that I enjoy on my patch; I find that walking the same paths means that you regularly meet the same people. Most of them are not people you would normally know or have much in common with. Some are dog walkers, some are joggers and occasionally you bump into other birders. I feel privileged to have met people belonging to much older generations who have an incredible insight into the changes that have occurred in the local bird populations and habitats. It is easy to make friends with other local birders as we share a common interest. You can give out information and tips and talk about interesting sightings in the local area, which can be great if you are not in the local social media-based bird groups. The very act of low-carbon birding can also facilitate new relationships. Not only do you meet the birders in your target area, but you can also meet people on your journey there and back. For some people, the act of journeying to their chosen nature spot can be as enjoyable as the spot itself—and that, I believe, embodies one of the special pleasures of low-carbon birding.

As birders we all need to consider our carbon footprint, and the more carbon intensive our lifestyle, the more we need to do so. There is an irony in the fact that the very birds that people make extensive journeys to tick off are often the rare migrants that have been set off course due to extreme weather events and changes in the climate. If you really feel the need to see that bird you could consider using public transport, or sharing a car with others. Or you may reflect on what is actually prompting that desire to see that bird. Perhaps you may find that the ultimate reasons have little to do with the enjoyment of the bird itself. And if patch birding or birding locally is not part of your birding routine, you may want to give it a try.

10

A Patch Year

David Raffle

Urban Newcastle may seem like a fairly unremarkable place to go birding, but this could not be further from the truth. A mere five kilometres from the city centre lies Gosforth Nature Reserve, which is about one square kilometre in size. It is a green oasis surrounded by busy roads, houses, industrial estate and farmland. Stepping into the reserve is like entering another world, leaving the hustle and bustle of the city behind to connect with the wilder side of the city. The habitats here are varied, featuring woodland, wetland and meadow, and support a whole host of wildlife. With over 1,600 species recorded, including 189 birds, there is always something new to discover.

I am lucky enough to have Gosforth Nature Reserve as my local patch, and have been making the short cycle ride there increasingly regularly over the last five years. Watching a local patch is a rewarding way to discover the wildlife on your doorstep and develop a stronger connection with your local area. Through volunteering, monitoring and watching the wildlife on the reserve, I have been able to get to know my local patch and notice all the changes happening around me. Tuning in to the seasonal changes on your local patch is captivating, and I find it to be one of the most fulfilling and enjoyable aspects of patch birding.

The brief days of the winter months are packed full of wildlife. The feeding station is a hive of activity: finches squabble over the food and tits make quick forays to the feeders, while blackbirds and robins hoover up the mess. Great spotted woodpeckers, nuthatches and treecreepers occasionally put in an appearance, flying in from the woodland and weaving their way around the bare tree trunks. Suddenly a sparrowhawk launches an attack and chaos ensues! Alarm calls ring out and the birds retreat to the bushes. Gradually, the birds return and the feeding frenzy resumes as if nothing had happened.

Moving further into the woods, roe deer run from the path as I approach, while a fox slinks away into the distance. Contact calls of loud parties of long-tailed tits can be heard as they fly between bushes, their distinctive lollipop-shaped outline silhouetted against the dull grey sky. Goldcrests tag along with tit flocks, foraging acrobatically in the treetops. The woodland opens up to alder, birch and willow, which dominate the wet margins of the reedbed. Flocks of siskin and lesser redpoll roam these areas, feasting on the abundance of alder and birch seeds.

You can hear the lake before you can see it; the sound of whistling wigeon and teal carries through the cold, still air and the reeds rustle as they sway in the breeze. The lake is full of wintering ducks as the resident mallards are joined by the likes of shovelers, pochards and tufted ducks, while goldeneyes and goosanders occasionally put in an appearance. In winter, I optimistically scan through the flock of wigeon and teal, hoping to find one of their American cousins, but without success.

The extensive reedbed is one of the best places to look for wintering bittern in north-east England. I will always remember my first bittern sighting, the excitement as it flew in and stood in the open, before its buff-and-brown plumage blended in to the reedbed as it stalked away. I have since spent many happy hours watching and waiting for these elusive reedbed denizens to appear. Sadly, the days when they were described as 'common as muck' by photographers at the reserve are long gone; now every sighting is to be treasured. While it is unknown why they are becoming less frequent here, it is possible that warmer winters in their breeding grounds have reduced the need for them to migrate to milder winter climes. Each autumn I wait in eager anticipation for them to return, only to be disappointed—their presence limited to a handful of fleeting sightings reported by others across the winter.

Snowdrops, primroses and daffodils add a touch of colour as the days lengthen and spring arrives. Queen bumblebees fly low over the woodland floor searching for nest sites. Frogs and toads return to the ponds, filling them with spawn. The breeding season is kicking off for birds too. Birdsong fills the woods as the songs of robins and thrushes are accompanied by the chiff-chaffing of the first returning chiffchaffs and the yaffle of the green woodpecker. Buzzards and sparrowhawks circle and display over the woodland. The heronry is busy with grey herons adding to their nests. Little egrets are an increasingly regular sight; how long will it be before they too start to breed in the reserve?

April is one of my favourite months on my patch—full of anticipation of the arriving summer migrants and birds on passage. There are often a few surprises thrown in too. A particularly memorable April saw garganeys, marsh harriers and otters all in the space of a week—not bad for a small urban reserve. The beautiful melodies of recently arrived blackcaps and willow warblers can be heard in the woods, towards the end of the month joined by the chuntering songs of reed warblers and sedge warblers emanating from the reeds. The number of swallows, sand martins and house martins hawking low over the lake increases by the day as more and more complete their long migration from sub-Saharan Africa. Common terns are next to arrive, fishing in the lake and preparing for the breeding season ahead.

May marks the return of the swifts, scything through the skies on their sickle-shaped wings in pursuit of aerial insects. The common terns focus their attention on breeding, laying their eggs in a shallow shingle scrape on the nesting platform provided for them. Regular visits are rewarded with seeing the eggs hatch and chicks grow, provisioned with freshly caught fish by the adults. Unfortunately, the last two years, 2020 and 2021, were sad ones for the terns, with all the nests predated overnight. The reserve's flora is just as rich as its birdlife. The hawthorn and blackthorn hedges are painted white with blossom, and the meadow is awash with flowers. The coralroot orchids, the second-largest colony in England, are also at their best.

As spring turns to summer, insect life reaches its peak. The meadow is buzzing with bees and butterflies, while dragonflies and damselflies are in abundance around the ponds. Four-spotted and broad-bodied chasers make short aerial pursuits before returning to their favoured perch, while emperor dragonflies patrol the wetlands. A wide range of butterflies can be found on sunny days, including recent colonisers of the reserve: dingy skipper and small heath. Climate change is thought to have driven the northwards shift in the range of several insect species in the last few decades, with commas, speckled woods, broad-bodied chasers and emperor dragonflies all becoming a common sight here and throughout wider north-east England.

The breeding season is in full swing, with chicks appearing all over the reserve. Pig-like squeals coming from the reeds give away the presence of water rails. Occasionally they put in a brief appearance, dashing between stands of reeds. Several pairs of water rail breed, and with some luck they can be seen with their chicks, little black fluffballs with long legs and a white bill, in tow. Gadwall and mallard ducklings, as well as little grebe chicks, can also be spotted on the lake. The mute swans and their cygnets are another highlight at this time of year. My favourite bird here is red PPU, a colour-ringed swan older than I am. She has been the resident breeding female the whole time I have been visiting, until 2021 when she was forced off the reserve by a new pair. The ability to observe bird behaviour and develop a connection to and knowledge of individual birds is one of the things that makes patch birding so special for me.

A different range of insects can be found as summer progresses. Warm, sunny evenings spent gazing up at the canopy of oak trees can produce purple hairstreaks, while checking the wych elms may be rewarded with sightings of white-letter hairstreaks. The meadow is visited by zippy fork-tailed flower-bees and Willoughby's leafcutter bee, while male bumblebee and cuckoo bees are abundant on the knapweed. Dragonflies are everywhere, with hawkers patrolling the ponds while darters rest on the paths.

Kingfishers perch in front of the hide to the delight of the photographers, posing like an electric blue and orange avian celebrity in front of the paparazzi. The summer of 2020 saw barn owls return to breed following an absence of over 40 years, with the ringed female coming all the way from Cambridgeshire. Watching one of these ghost-like owls hunting in semi-darkness after most visitors had left has to be one of my most memorable experiences at the reserve. Passage waders start to drop in at this time of year, signalling the start of autumn for many birders. Only small numbers of waders move through each year, but watching a black-tailed godwit, greenshank, green sandpiper or common sandpiper is so much more special on patch.

As autumn comes around, the leaves start to turn and insect life starts to shut down as the temperature drops. The woodland is a riot of colour: the trees are adorned with reds, oranges and yellows. The ground is carpeted with crisp leaves, rustling as jays rummage through them hunting for acorns to cache away for the winter. A woodcock flushes from the path, flying away through the trees and disappearing once more. Favourable north-easterly winds bring great flocks of redwings or fieldfares down from Scandinavia for the winter. *Tseep* and *chack* calls overhead make me look up in time to see great flocks of these winter thrushes appearing from the cloud overhead. Some drop in to the reserve but others continue to spend the winter elsewhere. This time I gaze up to find a skein of pink-footed geese heading south, probably from Iceland, as *wink-wink* calls rain down from above.

By November, most birds returning for the winter have arrived and attention turns to a true winter spectacle. The boardwalk is packed with crowds of people, all waiting in eager anticipation of the show that is about to unfold. The first birds fly in, soon to be joined by more and more until the sky is full of starlings. The starling murmuration truly is a spectacle to behold, with thousands upon thousands of birds all twisting and turning in synchrony in an incredible aerial ballet. The sound of their wingbeats and the gush of wind they create are audible as they pass low over the reeds before swooping back up into the sky. Suddenly, as if at the flick of a switch, the birds dive down into the reedbed. The starlings chatter and bathe at the edge of the reeds, preparing to roost overnight. But this huge gathering does not go unnoticed by the locals. Sparrowhawks wait until they settle in for the night, before racing through the reeds and sending up clouds of starlings in pursuit of their dinner. The sparrowhawk emerges from the mayhem with a starling dangling from its talons—success! As the light starts to fade, raucous flocks of jackdaws and carrion crows gather in the fields before wheeling around the treetops and going in to roost as the sun sets on another cold, dark winter's day.

11

Hunting Hawfinch

Steve Gale

My first hawfinch was behind glass, on show at a provincial museum—the work of a taxidermist. My curiosity for the species had already been aroused by illustrations and photographs in books. The stuffed bird before me looked back with dead glass eyes, apologetic of its faded feathers and imprisoned status. While once it flew, it was now consigned to sharing an airless space with other dust-filled bodies. Nevertheless, it was still an arresting beast. It was bigger than I had expected, with the heavy bill living up to prior hopes, a grey shawl and thorn-tipped flight feathers. I was to wait another couple of years before I saw my first live one, flying across a New Forest glade. There followed further meetings, most notably at a mid-Kent roost and a short-lived flock of over a hundred in a wooded North Downs valley. But this bulky finch would only reveal a bit of itself, rationing my experiences of it to a brief time or in the far distance. The winter of 2017–18 was to change all of that.

The events of that winter are already widely documented and have gone down in legend. Abnormal numbers of hawfinches were recorded across the country from mid-October, and although these movements had fizzled out somewhat by late November, pockets of birds were still to be found by diligent searching. I spent quite a bit of time looking for them, close to home, in places that I had recorded them before. My success was welcome, but of modest proportions. But modesty was about to be thrown out of the window as the month of January closed and February was ushered in.

I visited Juniper Top, in the Surrey North Downs, in the early morning of 30 January. It was not long before I heard a hawfinch and, as I wandered into the undergrowth, it was clear that the vocals were getting stronger and more numerous. Occasional shapes were moving through the treetops, soon morphing into hawfinches. The calls were now surrounding me and coming from directly above. They were continual, and for such a weak call, quite loud, betraying their proximity and number. Both the expected *tick* and *sip* calls were being made, but also a suite of unfamiliar noises that could be likened to a discordant electronic twittering. In front of me was a stand of mature, winter-bare beech trees, with maybe half a dozen yews dotted among them. It became apparent that quite a few of the hawfinches were regularly visiting these yews, and the comings and goings of the birds were constant. Any scan of the treetops provided me with 10–20 hawfinches, although accurate counting was hampered by the density and height of the twig-choked canopy. Small congregations would form, take to the air, circle round and land nearby, then make their way into the yews. The largest grouping reached 45 and there may have been at least 70 birds in the area. These busy birds took little notice, allowing me to wander in amongst them. After an hour they melted away.

The following morning, I entered the woods and returned to where most of the previous day's birds had performed so well. Although there was little calling, I located a tight group of 60 birds sitting passively in the tops of several beech trees. They slowly started to drop down into neighbouring yews, then proceeded to move away. By this time they were calling frequently and were easy to follow. A stand of larches some 100 metres further on had attracted the flock, and now it became obvious that the original 60 had joined others. The calling became incessant, a white-noise of *ticks* and *sips*—it was a frenzy. I stood underneath the tall conifers and watched as the birds moved further into the woods. To obtain a meaningful count I needed to be able to get a better viewpoint.

I lost the flock for maybe ten minutes but relocated it some 200 metres further on. With the sun unhelpfully in my eyes and wanting to get on the leading edge of the flock, I skirted around the birds and hid at the edge of a clearing. My timing was ideal and the first birds started to appear, moving through the canopy not unlike a tit flock. This enabled an accurate count as they moved past me—groups of 10–20, singles, one clot of 40—careful not to recount any bird that might double back. After 80 had moved through I became a touch excited, then onto the magic 100! But still they came, moving directly overhead and to the left, heading deeper into the wood. It was then that a single flock of 35 announced itself, having been hidden further down the eastern slope, and joined the magnificent mothership of hawfinches. By now the noise was at its height. I was experiencing a total immersion. The flock slowly moved away, dissolving into the woodland. Regardless of the accuracy of the count, there was no rest from the 'hawfinch hunt' as birds were continuing to call from where I had started, and I located a further 30 birds. These were certainly new. And, finally, after leaving those happily diving in and out of yew trees, a further flock of 40–50 birds appeared on the edge of the wood, at the very northern end of Juniper Top. I settled on a total count of 200 birds but I suspected that there were more. I needed to prove that there were.

For such a large passerine, the hawfinch is a surprisingly difficult species to observe and count. They are shy birds and have an almost supernatural habit of just disappearing in front of your very eyes, regardless of the number present. As I was discovering, with patience they could be tracked down and accurately counted. The beauty of this nationally notable gathering was that it was taking place on my doorstep and I had the time to devote to trying to find an answer to the question of just how many hawfinches were present. Back home, I spread out an Ordnance Survey map. Where could there be more? Was there a better place to clearly observe them? My time in the field increased and several sites were targeted. The Juniper Top birds that I had been watching tended to leave the area by late morning, so where did they go? I identified Bramblehall Wood as a place worth investigating, close to the other sightings and a bit off the beaten track. I was hopeful that the wood would be undisturbed, as there was no public footpath through it. All observation needed to be carried out from the edge of a horse gallops, looking across to the wood in question. My first visit was rewarded when I found hawfinches, and each time I returned I found more. Different times of day yielded vastly differing counts, with early morning by far the best time to visit. Throughout February and early March these totals grew into historically high counts for the country: 140 became 170 then 260. The realms of astonishment starting to be entered when 300, then 420, were realised. But there were still more to come.

Visiting Bramblehall Wood became an almost daily communion. Such an ornithological spectacle was once-in-a-lifetime material, the stuff of dreams. I was able to get up close to them, to watch them for hours on end, to get intimate with them—and all this on my extended patch. I could not get enough. 'Flock forays' were observed, when sizeable groups of birds would leave the woodland and fly out over surrounding open land in large arcs, returning to the area from where the flight started. This was often accompanied by frenzied, excitable calling, with chasing being an important element of these sorties. Pair bonding, with much chasing, bill-touching, and wing-flicking was observed. Some would take up prominent perches (normally at the very top of a tree) and stay in place for up to ten minutes, then fly out and make a short circular flight, returning to the same spot. Time was spent collating a list of their food choices: the seeds of yew, blackthorn, and rosehips; the fresh growth at the tips of yew branches; larch cones, and various tree buds. I was privy to not just awe-inspiring sights, but also sounds, one being when a sudden and loud 'whoosh' sounded above my head while I was standing underneath a yew tree—similar to the noise a starling murmuration makes when changing direction—and I was aware of a dark blur in front of me, and the air being sucked from around my head. It was a flock of around 150 hawfinches, spooked from the vegetation above me.

It peaked on 13 March. I was on site at dawn and was surprised to see 200 hawfinches already on show on the woodland edge. They were quite motionless and, I think it safe to assume, had just emerged from a roosting site very close by. Over the following hour more birds arrived to join them. From time to time some left the treetops to dive into the wall of yew beneath, barging through the dense foliage. A flock of around 100 then took to the air and headed purposefully northwards along the treeline, disappearing into the distance, not to be seen again. Then things started to get very busy indeed, beginning with a loose flock of 200 birds that came in over my head and circled those that were already present in front of me. This encouraged those in the trees to also take to the air—not the 100 or more that I had assumed were present, but at least 250 of them—and I was witness to a kaleidoscope of hawfinches, a blizzard of wing-bars, tail-tips, and excited calling, around 450 birds in all. And there were still around 50 birds behind me, up in the yews. Together with the 100 that had left northwards earlier in the morning that made for a minimum of 600. Within a minute they all disappeared, gone with barely a whimper, to be consumed by that dense wall of yew. All became very quiet indeed. Shocked is perhaps too strong a word to use to describe my reaction to all that I had witnessed, but it had certainly shaken me, in a positive and exhilarating way.

Bramblehall Wood did not take up all my 'hawfinch hunting' time. I scoured the wooded valleys within a three-mile radius of the Mole Gap, where the River Mole had worn through the North Downs chalk. The topography allowed me to look along clear sightlines, to identify distant stands of yew and further investigate. On many occasions such reconnaissance sessions resulted in flocks of hawfinches being clearly visible in treetops—I would take a slow walk towards them and often be rewarded with more birds before I reached them. I found further large flocks, particularly in the Ranmore area. Between there and Bramblehall Wood there were clearly over a thousand hawfinches.

The final signs of the irruption were played out in mid-April, leaving behind just a handful of breeding pairs. It had been a truly remarkable few months. It was made even more personal for me because the white-hot centre of hawfinch activity had taken place so close to home, in places that I already knew. That an ornithological happening of national note had occurred without the need for distant travel was, of course, down to luck, but it also prompted several questions.

Just how much is missed by the fact that many birders rely on getting in the car to drive miles to get their birding fix? If we were all to take a longer look, closer to home, just what would be discovered? My hawfinch hunting led me to explore areas close to home that I had not visited before and opened up other opportunities and discovery. Such discoveries might not have been of a thousand hawfinches, but that episode did prove that a major event, close to the edge of London, could easily have gone unnoticed.

12

In Praise of 'Projects'

Mark Bannister

Suddenly finding myself with more time on my hands, courtesy of a Brexit-induced early retirement in 2016, I decided to join a bird alert service and drive around to see rarities, as many do, if only limited to my local counties. This proved to be quite fortuitous timing, as Spurn recorded an exceptional autumn with seemingly endless Siberian vagrants. Maybe this is what they meant by a Brexit dividend? Parallel to this, having been inspired by Ian Newton's New Naturalist *Bird Populations* and studies by local birders, I decided to also try a few of my own locally based 'projects'. I subsequently found that the local projects generated by far the most enjoyment and have since become the focus for my birding while the driving-around-for-rarities has withered to near non-existence, apart from trips to the east coast in autumn. Perhaps the main reason for this is that my interests have always leaned towards a better understanding of bird behaviour, their lives and populations and, in addition, I do like data—most likely this is the ex-engineer in me, having spent half my life analysing and plotting 'stuff'. The rare bird alert subscription has been binned and I do not miss it.

The growth in my time spent on local projects also neatly dovetailed into a desire to replace journeys by car with the bicycle wherever possible in order to minimise my carbon emissions. I have been using a bike to bird the estuary and the Humber bank clay pits on the doorstep of my home town of Barton-upon-Humber for over 30 years, mainly for convenience—most of the pits lie on a nine-kilometre stretch to New Holland, which I find just a touch too far to walk for my time-limited birding sessions. I am lucky to have it on my doorstep: the extensive reedbeds and scrub still hold abundant breeding warblers (up to 780 territories of nine species—data from my first project!), the mass hatching of insects draws in large flocks of hirundines, and flocks of swift up to a thousand strong, driven down by poor weather, are a spring highlight. Bittern, marsh harrier, bearded tit and passage migrants are present along with regular rarities. The decision to use the bike instead of the car for projects further afield was therefore more of an incremental step. If my back is not playing up, I rarely drive, now, to local sites less than 24 kilometres away. It has surprised me how little of a problem this has been, and I am convinced that I actually see more birds than I would when using the car.

My main ongoing project has been to monitor the breeding success of waterfowl on the local clay pits. While the 25 clay pits I study are nominally similar and experience the same weather and diversity of predator species, they vary to a degree in age, size, depth, open water extent, water quality, public access and human activities. Although not unique, this provides a rare opportunity to examine what influences breeding success. Every pit surveyed is visited by bike, at least twice a week, in a 'constant effort' fashion. I decided against relying solely on numbers of breeding birds present as this may be influenced by immigration/emigration and the presence of non-breeding birds, so I also obtain an estimate of productivity by following the fortunes of all broods I manage to find each year—averaging around 140. I have still not tired of the thrill of discovering yet another brood, even if only a moorhen. New life is always quite magical to witness. Although I have never really taken part in 'listing', a garganey or shoveler brood would be nice—perhaps my own equivalent of finding an American vagrant for a UK list—but I am still waiting!

Combining the British Trust for Ornithology (BTO) survival data with the productivity data from my own study, the resulting implied population changes roughly agree with the declines shown by the BTO population indices for the East Midlands. Coot, moorhen, all the duck species and great crested grebe all appear to be producing too few fledged young to sustain the local populations, and they rely to some extent on local immigration to help sustain numbers. Little grebe is just about stable, while mute swan and the feral geese (Canada, greylag) are strongly increasing.

Teasing out what really matters to the bird populations from the data is of course tricky, but some things are obvious. Firstly, the top seven most productive pits all have very high densities of aquatic plants (macrophytes) to the extent that they are visible on the surface over large areas. Many duck species will nest in one pit and move their young to another that is macrophyte dominated but less suitable for nesting. The availability of food which can easily be accessed by young waterfowl close to the surface appears to be the main attraction. The loss of several of these local macrophyte-dominated pits to dye application for water sports, agricultural run-off and nutrient load from the feral goose flocks is one of the main negative impacts on productivity. Another reason appears to be an increase in both the number and diversity of predators, native and non-native: mink, otter, three species of gull, marsh harrier and bittern are all either new or have arrived in greater numbers since the late 1990s, which may explain the current trend of larger birds being more productive, as they are more able to fend off predation.

Another key change over the last 30 years is that the emergent reedbed areas favoured for nesting are being lost to grazing pressure by the increasingly large Humber feral goose population. Great crested grebe has lost several of its old nest sites in this manner and is now limited to regular breeding in just a few pits. It is likely that coot, little grebe and pochard have also been similarly affected. This slow reedbed loss is easily missed due to 'shifting baseline syndrome', that is, our inability to recognise long-term environmental change due to our tendency to assess change based on our own biographical experience rather than on long-term historical information. However, when comparing images over the years using Google Earth, it is clear some pits have lost over 30% of reedbed and a far higher proportion of the emergent reedbed as it is grazed from the water.

Surprising to some, perhaps, is the fact that I can find no impact of visitor disturbance on the breeding success of any of the species studied. Five of the seven pits with the highest productivity are actually within the country park and receive at least an order of magnitude higher footfall than elsewhere, including many dog walkers. It is known that disturbance to birds is proportional to their distance from a safe refuge, and it appears likely that the species I study regard the local wetland habitat as just that (in contrast to, say, a ground-nesting bird on a beach), resulting in no detectable effect at the population level.

As the data builds over the years, it has become clear that weather events have the biggest effect on productivity. The number of fledged young follows a cyclic year-to-year fashion, described by Newton as being due to predator–prey relationships, but I have also found that the poor years have corresponded to low levels of local rainfall in spring. In a good year over 300 fledged young of all species can be produced (excluding feral geese), but in a bad year, this can easily be halved. The same effect has been found on the

North American prairie potholes. During these dry years, I have noted far more predation, likely due to easier access to nests by terrestrial predators and increased avian predation, perhaps due to a lack of alternative food sources from the hardened ground such as earthworms. Climate scientists have noted that if average global temperatures rise by 2.0–3.5°C, mean summer temperatures in the east of England may rise by 5°C, and although winter rainfall is projected to increase, summer rainfall may decline by 50%. It is not clear what will happen in the main spring–midsummer nesting period but if drier weather within this period becomes the norm then poor breeding years are highly likely to become more frequent. This would appear to be by far the biggest threat to the bird populations studied.

My bird 'projects' are, of course, nothing new. Many do similar studies as part of their own birding, and to a far higher standard than mine; but whatever your standard of birding, there is always something to learn and enjoy. Perhaps there is scope for all of us to focus our attention on specific aspects of the natural world and find out things that, if not ground breaking, are not widely appreciated.

After a break due to lockdown during the Covid-19 pandemic, I am looking forward to continuing my corn bunting survey. Cycling the minor roads and bridleways over the chalk wold to Grasby is a rather lovely way to spend a spring day. I am hopeful that our remaining relict population of 70 singing males has increased. With ash die-back on the near horizon, I also really need to add tree sparrow to this survey, as our local population on the wold appears to be evenly split between hole nesters in old ash and those under the pantiles of farm buildings. Perhaps there will be an increase in nesting opportunities in the dying ash before a crash in the population as the trees are lost from the landscape, a little like what happened with lesser spotted woodpecker and the elms in the 1970s? I also need to make a better attempt at surveying raptors on the wold and in local woods. With a bit of luck, goshawk will be with us soon as a nesting species which may shake things up a bit. The buzzards have had it their own way for far too long. On the other hand, I keep putting off a town swift survey: which buildings hold breeding swifts and are they decreasing? So many projects and so little time. I do not think I will be running out of them any time soon.

13

The Backyard Jungle

Finley Hutchinson

Picture this: an urban metropolis, stretching as far as the eye can see, bathed in the dull grey tones of modern living and intersected by rivers of cars cutting through the landscape. Bustling with day-to-day life and yet seemingly a desert for nature and animals. It is all too easy to jump in a car, or on a plane, and trade this hostile environment for something more pleasingly natural where Mother Nature rules the landscape. And sure, from time to time it is nice to get away, but in our carbon-guzzling society it is essential—now more than ever—that we start to appreciate the inconspicuous gems that are, and have always been, hiding in plain sight on our doorsteps. We cannot afford to keep on spewing out carbon dioxide at today's rate, and there is a whole world under our noses that can be just as satisfying as high-carbon pursuits—perhaps even more so—to explore and get to know.

I have grown up in the city, surrounded by noise and pollution. Despite this, I have always been exposed to nature—thanks largely to my parents' love of walking and my grandparents' birdwatching. I have been on the lookout for nature from a young age and have been birdwatching for half a decade. But it is only in the last couple of years that I have really got to know the wildlife on my doorstep.

Much of this has been due to the Covid-19 pandemic. Being a student in Year 11 (GCSE year), when exams were cancelled and school was suspended, I was left with a huge amount of free time with no other commitments, although I was largely confined to my house and garden. I used most of this time, from the start of lockdown in late March 2020 right up until I finally got out of the county for the first time in August 2020, to get to know and appreciate the nature right outside my backdoor.

If I had to pick just one moment, then I would say that my new-found passion for entomology came from an event in the first week of the first lockdown. It was nothing major—just the simple act of emptying out the detritus that had accumulated at the bottom of an old bag of logs. Yet I was astonished by the huge array of invertebrates found within: my first springtails, several different species of woodlice, a hairy cellar beetle, a spotted snake millipede, an assortment of soil mites, tiny translucent nematodes waving their bodies in the soil. I spent over an hour lying on my front with my camera and macro lens, photographing the diversity scattered before me, and then a long time on top of that processing the images and identifying what I could. Almost all of it was new to me; not because any of it was uncommon, just because I had simply never opened my eyes and *looked* for it before. From that moment I knew I was hooked on entomology, but I still could not have predicted how far it would take me in such a short amount of time.

Very soon after this I got my first microscope—a Brunel MX5t. I found it incredible to see all the tiny details that I had not been able to appreciate before. A daily pattern developed during the next few months of lockdown: go outdoors, find an insect, bring it back to identify it with the microscope, release it and find another. This was interspersed with regular garden birding sessions, including two 24-hour 'garden bird races' that I undertook with friends—it is incredible how much flies over a medium-sized urban garden that is not really on any major migration route. I saw or heard many unexpected species during the first lockdown: hen harrier, common scoter, whimbrel, crossbill, dunlin, little ringed plover… Coots and moorhens routinely flew around overhead after dark—before listening out at night I had no idea that they ever ventured from the local lake. A few times, I stepped outside in the evening and heard the evocative whistling calls of wigeon migrating over the city—probably my favourite bird call and one that never fails to lift my spirits.

In January 2020 I received a moth trap for my birthday, but I did not catch anything until lockdown because until then I did not have the time to properly put it to use. By trapping every other night over the summer, in one year I managed to build up a garden moth list of over 380 species—another thing that I would not have thought possible in an urban area before I gave it a go! Both quality and quantity have been much better than anticipated—I will not forget catching the fourteenth UK record of the micro-moth *Cydia interscindana* or the first Berkshire record of light crimson underwing this century, although the experience of more than 500 water veneers flopping feebly around the trap on a warm summer evening is not one of my fondest memories! One of the best things about moth trapping is how easy it is to uncover a whole new world that is hidden by day; moth traps are available from most naturalist supply shops and can even be made with as little as a torch and a white sheet. Some words of warning though: once you start, there is no going back!

Of course, moths are just a drop in the vast ocean of invertebrate life hiding just on the other side of the window. There is endless variety and beauty waiting to astonish us, and over a year into my entomological journey I am still regularly amazed by what I find. You could spend an entire lifetime studying invertebrates and you would still experience unimaginable things out of the blue. That was my thought when I found a 0.6-millimetre-long wasp in the garden; when a new species for the UK arrived at my moth trap; when I watched as a tiny parasitic wasp emerged from a parasitised scale insect before my eyes… What will come next?! It is this anticipation that has kept me interested and I am sure will continue to do so for a long time.

One group of invertebrates that I find particularly fascinating are the Collembola, or springtails. Before lockdown in March 2020 I had never knowingly seen a springtail, but once I started looking out for them I realised that they are quite literally everywhere! I first saw one of the UK's largest species, *Orchesella villosa*, at around 4 millimetres long, and have now seen over 50 species, although I am still searching for the 0.5-millimetre *Megalothorax minimus*. Springtails have six legs and yet they are not insects, but rather their own separate lineage dating back at least 412 million years— that is nearly twice as old as the first dinosaurs! Springtails can easily be found jumping away when you move a pile of leaf litter, but they occur in almost every habitat, and they are so understudied that you can make amazing discoveries wherever you are.

I cannot write about my interest in invertebrates without mentioning the much-maligned wasp. Anybody who knows me or follows me on social media will likely have realised that I am a big fan of wasps—something which surprises many people who immediately picture the black-and-yellow stinging things that terrorise picnics on summer days. And yes, I do like them too—but I am much more interested in the wasps that most people will see and say 'oh, look, a fly'… or not notice at all! Before I became interested in entomology, I had the same perception of wasps as most people do, but I have found out so much about their diversity and biology in the last year, and I continue to be amazed by them on a daily basis. In the UK there are over 7,000 recorded wasp species, of which only a small handful sting people, and there are so many more waiting to be discovered. The vast majority are parasitoids, laying their eggs in or on other organisms, which the larvae then consume upon hatching. A large proportion of the total number is made up of the ichneumonids (around 2,400 UK species, making up 10% of all the UK's insects!) and the tiny chalcidoids (comprising over 1,700 UK species, ranging from a few millimetres in length down to just 0.3 millimetres in the case of the mymarid (or fairy-fly) *Alaptus minimus*). Spider-hunting wasps, numbering 44 species in the UK, can often be seen on warm days flying and running over vegetation, seeking out spiders which they paralyse and carry back to their burrows for their young to eat alive when they hatch. Many wasps are even 'hyperparasites' of other wasps, laying their egg on a wasp larva that is itself a parasite of something else—the record number of layers of hyperparasitism is seven! That is a wasp larva feeding on a wasp larva feeding on a wasp larva feeding on a wasp larva... and so on. Like the springtails—probably even more so—wasps are immensely under-recorded. Even in the UK, where invertebrates are often much better recorded than in other countries, lots of species are known from just a few specimens. Anybody who chooses to take an interest in this amazing group of insects can make a significant contribution to science, no matter where they are—my urban garden has already produced a new wasp species for Britain!

As I write this the UK is preparing to come out of lockdown once more—hopefully, for the last time. For many of us, this will mean being able to travel much more freely again, and I am definitely looking forward to being able to go slightly further afield. But, at the same time, my eyes have been opened to the astounding variety of nature that can be found close to home, and there will always be time for a good bit of garden birding or entomology. Hopefully, this chapter will help others to realise that too and share in the beauty of our backyard jungle—as well as help to protect it for generations to come. Not only will you be blown away every single day by what you find, but it also works wonders for your mental health.

14

My Patch and the Plastic Problem

Siân Mercer

Plastic pollution is rapidly becoming one of the most concerning environmental issues of our time. The production of plastics is increasing exponentially, and research suggests it is even expected to double by 2050. The flexible, durable and versatile properties of plastics have led to them being incorporated into almost every element of our lives and they have gained the status of a 'wonder' product and an essential in modern day living. Plastic can be found on nearly every supermarket shelf, in our online purchases, clothes and even in skincare products. We are living life surrounded by plastic, and in the last decade we have finally started to see how our consumerist attitudes have created this problem. A major culprit lies a little closer to home, in most of our driveways. Tyres. In 2017, the International Union for Conservation of Nature (IUCN) estimated that 28.3% of microplastics in the ocean come from tyres, landing them in the top seven contributors. Emissions from vehicles have been gaining international attention for their role in our climate crisis, but recent research suggests that this surprising source of plastic could pose a serious threat to us and our environment. Although we often refer to tyres as 'rubber' most of their composition is a mix of synthetic compounds such as fillers, binders and other chemical additives. Natural rubber—which is extracted from trees—only plays a minor part as a component, while synthetic polymers make up the majority. Once these are broken down, they can easily enter the environment. Producing tyres also requires burning large amounts of oil—7 gallons (32 litres) for a modern car tyre and 22 gallons (100 litres) for a truck tyre—hence contributing to climate change.

Continuous use of vehicles across hard, abrasive ground gradually wears away the surface of tyres until inevitably they need replacing—when roads are wet the tyre particles (microplastics) left behind are washed off into our waterways. These eventually lead to our surface water stores, such as lakes, rivers and oceans. A report commissioned by Friends of the Earth estimated that half a million tonnes of tyre-wear fragments are released every year across Europe, leading to potentially damaging and deadly consequences for our waterways. Once they have entered our ecosystems they can absorb and concentrate toxic pollutants from the water, making them more of a threat to wildlife through either ingestion or contact with skin and gills. Equally, they pose a threat to life on land—it is estimated that 10% of plastic from tyre wear becomes a form of air pollution, which is contributing to rising lung disorders that are related to airborne particles. Due to their size, it is difficult to fully assess local and worldwide effects of tyre particles on humans and wildlife but this continues to be the subject of research.

While plastic is such a useful material it also poses a tangible threat to our environment, as the long-lasting substances that it is made of can take up to 400 years to decompose. The harm caused to any habitat from chemical and microplastic leaching is inevitable. Microplastics, broken down by wind, waves, sea and sun from larger plastic debris, have been found across the globe, from Mount Everest all the way to the Mariana Trench. Tests have shown that in some species, such as oysters, the consumption of these substances and materials can cause significant damage to the liver, tissue and reproductive systems. Plastic microfibres have even been discovered in municipal drinking water systems and drifting through the air. Alongside this, millions of animals are killed by plastics every year—although most of the victims are found in our waterways and vast oceans, land-based mammals have been known to die from plastic consumption, left as litter outdoors. The true extent of the problem is difficult to measure and hard to comprehend, especially since we—humans—are, after all, the only culprits.

In our modern era it is a difficult balance to strike: the convenience and need for plastics in our lives is undeniable. Although no one individual can reduce the sheer scale of global plastic pollution, through consumer-driven campaigning and purposeful life choices there are positive actions we can all take, from small to more substantial changes. It is often difficult to live a completely sustainable life, but we should celebrate what efforts we do make instead of bemoaning what we cannot change. Small alterations to our work, school or personal lives can not only have positive effects on the environment but equally positive effects on how we feel—making a good choice for our planet is often a good choice for ourselves!

Over the Covid-19 lockdown, I was inspired by the work of amazing clean-up groups around the coastline who were tirelessly collecting the plastic they found washing up on their shores. Living in a rural, inland village in Shropshire I was far from the ocean but had the same determination to help wildlife where I lived. During my time at secondary school I had been passionate about the environment, doing projects on microplastics, raising awareness about recycling, holding an environmental protest and petition to encourage other pupils to understand the consequences of our actions on the Earth. I also participated in setting up a 'refill' scheme in my local town to advocate against the use of unnecessary plastic bottles when possible and instead focus on using reusable, and refillable, water bottles. During 2021 I began noticing a significant increase in litter on my lanes; as a result I began picking up pieces of rubbish whenever I was out, before realising the problem needed a more tactical approach. Armed with bags, gloves and a litter picker (for sanitary reasons,) I managed to find over 200 pieces of litter within a short two-mile walk. The contents included disposable masks, balloons, sunglasses, wipes, food packaging, plastic bottles, cans, electrical wires and tyres. I was shocked to see a wide array of products simply discarded and disposed of in such an irresponsible and selfish manner. Some dates were still visible on packaging, one reading '2008'. When I walked around other parts of my village I recognised an alarming pattern and rubbish in some areas almost had a form of normality, especially empty cans of alcohol in ditches. Even walking around my patch, a usually quiet reserve which attracts walkers and wildlife enthusiasts alike, I noticed wipes and masks discarded on paths. An important part of patch birding for me is loving the place in which I live—by tidying up the rubbish we can encourage nature to flourish and hopefully inspire that same love in someone else.

Incorporating litter picking into our daily strolls, walks or jogs in and around the patch is a great way to stay active but also deal with any rubbish before it becomes a problem. Preventing pollution in your local area may not be a large gesture but turning up and playing your part is enough to have a small-scale positive impact. From platforms such as Twitter, I have heard of many brilliant and innovative ways to integrate such action into day-to-day life: during the school run, while walking the dog and even as part of the activities of local community groups such as scouts or youth groups.

Efforts to reduce single-use plastics in general are not only beneficial to where you live but to the rest of the world too; using metal straws, reusable bottles and cups, bags for life and avoiding plastic cutlery are all low-level swaps which are achievable for the majority of people. These actions help reduce the presence of macroplastics in our local areas. Microplastics are far trickier, due to their widespread distribution and small size.

There are opportunities for us to make better choices in many areas of our daily lives, though. Reducing use of our cars and switching to greener modes of travel such as bikes, walking or public transport play a massive part in this. Over lockdown I enjoyed exploring both my local lanes and patch by bike, allowing me to experience nature while helping to reduce the output of microplastics into the environment. Where car journeys are unavoidable, drivers can help by adopting techniques known as 'eco-driving', including accelerating gently, driving with the correct tyre pressure and avoiding abrupt and aggressive manoeuvres. These all reduce the quantities of plastic shed by tyres. Making choices based on the component list of products, for example in cosmetics and skincare, is another thing we can all implement into our routine in the same way as reducing our consumption of single-use plastic packaging. Moving away from easier plastic options can create consumer-based change and, when enough of us do this, encourage companies to think about their sustainability and effect on the environment. We, as citizens, can campaign to ask governments to regulate both the production and disposal of plastic so that these companies are forced to act by law. With massive budgets and new research they have the power to create significant global change. Plastic pollution is not just a problem of the past and present but also of our future. Influencers around the world continue to persuade a new generation of climate-conscious children to make better decisions and care about our planet. Young people play an important part in this fight and engaging them through documentaries, social media and campaigns is proving surprisingly effective. It is safe to say that, with the talented individuals that tirelessly work for progress we, along with our wildlife, are in good hands.

Birdwatching will always go hand in hand with my passion for the environment and the threats it faces. The beauty and tranquillity I find in nature is unparalleled and saving it will always be close to my heart. I aim to continue to clean up my local area alongside learning about the wildlife that calls it home. It is important to me—now more than ever—to appreciate the value of where I live and express how grateful I am for it.

15

Eleventh-Hour Birding

Simon Gillings

For much of my birding life I have been envious of people who have a high-quality local patch on their doorstep—somewhere close enough to visit frequently and get stuck into understanding the comings and goings of different species, and maybe find the odd thing that is scarce in the area. There were some superb gravel pits on the outskirts of the Lincolnshire village where I grew up, but since leaving home, I have struggled to find a suitable replacement. Rather unexpectedly, since early 2017 my local patch has become the dark air space above my garden.

In the mid-2010s birders began to hear of the astonishing numbers of ortolan buntings detected by sound recording as they migrated over Portland Bill, causing one of two reactions: sceptical raised eyebrows or an opening of minds to this new nocturnal frontier of birding. I cannot really remember which camp I was in at the start but to me, living in a Cambridge suburb, miles from the coast, this seemed like yet another birding opportunity that was out of my reach. Sound recording also seemed like a specialist and expensive hobby to dive into for what were likely to be slim pickings. Around this time our second child was born, precipitating the conversion of our third bedroom from office to nursery, and the displacement of the family work space to a shed in the garden. At some point I got a very cheap (around £10) USB mini-microphone for video calls and often posted this through a hole in the wall to listen to the outdoors while working in the shed. This was not initially for nocturnal recording but when my friend and British Trust for Ornithology (BTO) colleague Nick Moran recorded a bittern and multiple water rails over his garden in Thetford, I decided to give it a try. Rather than invest in loads of new kit I rigged up the mini-microphone in a Pyrex mixing bowl (in the misguided hope that it might act as a parabolic reflector), connected it to my computer, and waited. I did not need to wait long. On the evening of 15 March 2017 at three minutes past ten I recorded and heard first-hand the unmistakable *kek-kek-kek* call of a night-flying moorhen.

That moorhen was the first record for my garden and was quickly followed by the first water rail. There are moorhens a few hundred metres away on the River Cam but in ten years of living at the property I had never had a sniff of one over the garden (with hindsight I realise I had not been trying very hard). Water rail on the other hand felt mega: I had just one record, from along the river the previous winter, and the nearest breeding site some five kilometres away. Over the next few weeks moorhens flew over almost nightly, and I soon added wigeon, teal and other hitherto undetected species. I became hooked on sound recording as a means of detecting birds in this new-found patch—an inverted cone of airspace above my shed. I upgraded my microphone and started reading as much as I could about different bird flight calls, scouring xeno-canto (an online global repository for recordings of bird songs and calls) to learn from other recordists, especially Dutch counterparts who had been recording flight calls for several years. I learned how to read a spectrogram. This is a key skill because, when faced with ten hours of audio from a night, it is impractical to listen back in real time. Instead the digital audio files can be displayed as a spectrogram (also called a sonogram) which is a two-dimensional visual representation of the sound, with frequency on the vertical axis and time on the horizontal axis. Spectrograms can be produced using free software such as Audacity,[1] and by zooming in you can look at each consecutive 30-second snapshot. As with frames of an old film reel, you can page through these looking for interesting sound events. I developed a unique insight into the ambient soundscape of my immediate surroundings, for example learning what the sound of a squeaky bike looks like so I could ignore it when paging through the spectrogram. I discovered that my neighbour's shed door makes a sound like a black redstart and that foxes and tawny owls are much more regular visitors than my conventional observations would suggest.

My initial interest in recording nocturnal migration or 'nocmig' (I still wish 'invismig' had taken hold...) centred around adding species to my garden list but that developed into a more systematic endeavour.[2] I wanted to know how regularly these different species flew over my garden. I was particularly interested in the rails. Although this strange offshoot of birding had the word 'migration' in its name, surely the moorhens I was detecting in May were not still migrating? Having worked for the BTO for over 20 years I strongly adhere to the view that it is difficult to detect coherent patterns from bird records without measuring the amount of recording effort expended and being consistent in what you record. For nocmig recordings that means noting down the start and end times on a recording and then being consistent in logging every species. This 'complete list' ethos is the same as used by BirdTrack and other online platforms and translates equally well to nocturnal migration records: I can look back on my nocmig records and be confident that a gap or dip in the detections of water rail records indicates they were scarcer and not simply that I got bored of logging them or spent less time trying. I also try to record throughout the year and not only during the main migration periods, just to see what else is going on that we might have missed.

Since March 2017 I have made nocmig recordings in Cambridge on over 950 nights, totalling 5,700 hours of recording effort, and detected approximately 23,000 individuals of 85 species. By far the commonest are the thrushes, with redwing accounting for 67% of birds detected,

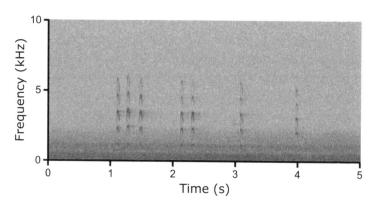

Figure 1 Spectrogram of a series of moorhen *kek* call notes. Cambridge, 22:38, 30/05/2020 (S. Gillings). See: https://www.xeno -canto.org/681705

followed by song thrush (11%) and blackbird (6%). But I have also detected 22 species of wader, over 100 water rails, 85 little grebes, 38 tree pipits, 19 Sandwich terns and many others. I did not start nocmig recording in a conscious bid to stay local or to reduce my birding-related carbon footprint but to some extent it has achieved both. Nocmig has not replaced birding but has complemented more traditional forms of birding. It has given me a way to stay connected with birds and migration at times when I cannot get out due to other commitments. Sure, I would rather see a whimbrel migrating in the flesh, but there is still something special about scrolling through a spectrogram and seeing the tell-tale signature of a whimbrel trill. Some birders are quite sniffy about nocmig, questioning its worth or asking how you can tick something you did not personally experience. I admit, at times it is like I am sharing my patch with another birder, someone with no commitments who can bird there every day and, without fail, messages me to tell me what I have missed. But the way I see it, these are still my discoveries, captured on my initiative, identified by my eyes and ears, even if after the fact. I see little difference from ticking moths caught in a moth trap: those

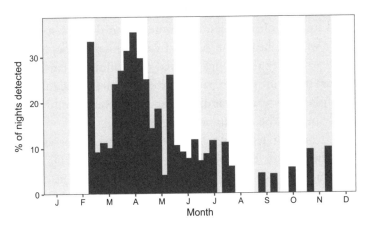

Figure 2 Percentage of recording nights per week when water rails were detected. Data from Cambridge for 2017–2020.

moths would not be there in the morning had the mother not left a trap running.

Garden ticks aside, nocmig recording has revealed to me the diversity of species that are on offer locally if I modify my birding behaviour and efforts to experience them firsthand. I now know that species as diverse as little grebe, common scoter, ring ouzel and tree pipit regularly migrate over my garden. Some of these are so rare that I would not stand much chance of hearing them by listening at the back door with a final cuppa before bed. But nocmig has taught me which nights are best to sit outside reading after dark, or, if I am working late on some new Alula design, when I should live-listen to the audio stream. Having previously recorded at least 79 common scoters and 8 turnstones, I heard both these species first-hand in 2021 because I knew when and what to listen for. However, one species still eludes me. My nocmig data show that Sandwich terns regularly migrate over Cambridge during September, usually in the early evening between nine in the evening and midnight. These are presumably birds leaving the Wash at dusk and heading cross country to the west coast. I know they often involve family groups as my recordings contain the typical scratchy calls of adults and the shrill squeaks of juveniles. For some reason this is a species I really want to hear from home. I came closest in September 2020 when my recordings showed one calling distantly as I walked down the garden path. I have not quite got to the stage of piping the nocmig audio stream into every room in the house but maybe it is just a matter of time.

So, nocmig recording has become part of the range of activities I think of as birding. In addition to learning about the true status of birds locally I have learned a considerable amount about bird flight calls that I can put into practice in the field. On two occasions, dotterels flying over would have been missed had I not known the call from my nocmig research. Furthermore, nocmig has allowed me to experience birds in distant places without the associated travel-related carbon emissions. I have a long-running recorder in my parents' garden in Norfolk and in 2020 I sent an AudioMoth recorder to my friend Dan Chamberlain in northern Italy. Dan hosted this small programmable recorder on his shed during 2020–1, giving my ears a chance to go on a foreign holiday even if I never left home. Those recordings gave me experience of night-herons, little bitterns and black-winged stilts; hopefully, useful preparation for when one flies over my own garden!

Now, in 2021, my birding is evolving again. Events over the last three years have made me seriously question my birding activities. In 2019 I attended a very enjoyable conference in Israel, shortly followed by a trip to Kenya to teach African scientists ecological and analytical skills, then later in the year a trip to Nigeria to teach students and discuss citizen science. Inevitably, I also saw some great birds on those trips but by the third trip I did not feel I could talk about the birds. Whereas normally I would share photos on social media on my return, this time I decided not to. Seeing stunning photos of African birds is what made me want to go. Better to not

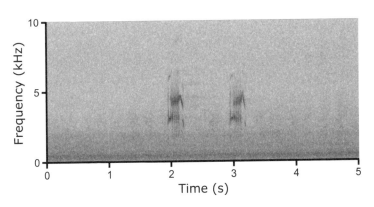

Figure 3 Spectrogram of a pair of Sandwich tern calls. Cambridge, 21:34, 22/09/2018 (S. Gillings). See: https://www.xeno-canto.org /681716

share mine and nudge others to repeat my damaging actions? But, in truth, I was probably also embarrassed that at a time like this I had flown to Africa not once but twice, even if the underlying motivations were well intentioned. Those trips, other events and a growing fear around climate change have hardened my resolve to limit my birding as much as possible to on foot and bike. I turned this into a personal challenge to see as many species as possible by bike in 2021 and so far have surprised myself at what I can see, and how far I can cycle. Perhaps the next step in this evolution is to combine sound recording and biking. When local birder Jon Heath found a great reed warbler just five kilometres from my home I quickly discovered how difficult it is to cycle with a parabolic reflector. Maybe the challenge for 2022 will be to see how many species I can sound record by bike.

16

Listening Again to Birdsong

Dave Langlois

Birdsong is the basis of all human music. No less. A big claim to start this short essay. But when you think about it, it must be true. When and wherever it was that the first human beings managed to produce a tune on some hollow reed or bone the only reference they would have had, the only environmental sound with any real melodic content, would have been birdsong. Birds, after all, have been at it for a few million years now. And the blackcap, robin or nightingale that served as role models for that first human tune would have sounded more or less the same as the ones we hear today.

But there is the thing. Few bother to listen anymore. Some years ago now, my son and I were dining in a Shetland restaurant and the owner told us that a pod of orcas once turned up in the bay right outside and the whole place emptied in seconds, except for one man who carried on imperturbably eating without lifting his eyes from his plate. With birdsong it always seems to me the other way about: I am the only one running out to witness the miracle. For miracle it is, without a shadow of a doubt. A miracle honed over those millions of years to serve the two main purposes of attracting the female and defending the territory.

But why on earth have these songs become so complex? Why such an astonishing range: the heart-melting mellifluousness of the blackbird, the staggering extravaganza of the nightingale, the effortless gush of the blackcap, the suicidal crescendo of the wren, the pell-mell gallop of the garden warbler, the stentorian iterations of the song thrush, the cheerful refrain of the chaffinch, the sobbing chromaticism of the woodlark, the tripping cadence of the willow warbler, the breathless fantasia of the skylark… ? Would not a simple *beep beep beep*, like a reversing forklift truck, have sufficed?

Well, there are several factors that have driven this complexity over these millions of years. Firstly, the purer the note, the more directional clues it gives, due to the staggered arrival of the sine wave at both ears. This is why, conversely, alarm calls are so discordant—in order to give no whereabouts clues to the predator. We have to remember here that birdsong is a sort of karaoke with the females sitting as implacable judges of the male's performance. Little would it suffice for the female to single out an outstanding male if she then lost track of him on the dancefloor. When she is incubating, too, she needs to know where *her* male is at all moments, and he needs to assure her constantly that he remains a crack performer otherwise she will not hesitate to cuckold him. DNA studies have shown that a very high proportion of nests have eggs fertilised by a male other than the one ostensibly protecting that nest.[1] In fact, only 10% of apparently monogamous pairs are in fact faithful. Infidelity is the norm. Birdsong is the antidote.

So much for tone, then: a pure sound, like that of the majority of songs we hear every day, will automatically have been selected for over the aeons. But what about structure? Well, the more complex the song, the more room there is for individual differentiation within the recognisable speciation. Songs are not innate. Or not completely. They have to be learned and the cognitive feat of learning them perfectly is what the females are listening out for in the karaoke. A male, after all, that has managed to learn the song well will probably be good at other cognitive feats like defending a territory, finding food, protecting the young or other daily deeds of derring-do. The males, therefore, need to reproduce the species' song faithfully but with enough idiosyncrasies (barely perceptible to our ears) to enable the female to identify her own particular mate. This differentiation saves the male no little effort too. There is a phenomenon known as the 'dear enemy' effect. If a male recognises another singer as a known neighbour, who is behaving itself in its own territory without any encroachment on the former's territory, there will be a sort of *entente cordiale* between them. If a male suddenly hears another unknown male, however, he is likely to react much more aggressively because this new invader, with nothing to lose, might prove to be a much bigger, all-for-nothing threat.

All this would be impossible with a simple *beep beep beep*. And it is this fine juggling act of reproducing perfectly the species' template, as a sign of its cleverness, while retaining, too, its own individual stamp, that has driven the evolution towards increasing complexity.

Another fascinating chapter in this whole story is that of mimicry. Prima facie, imitating other species would seem to run counter to the abovementioned tenet of perfectly reproducing the own-species' blueprint as a show of cognitive ability. But, for whatever reason, the females of certain species like starling, redstart and marsh warbler seem to accept a roster of imitations as an alternative sign of this same smartness. Until recently my star performer was a redstart I heard in Wales back in the 1970s, capable of imitating 14 other birdsongs. I always thought this was pretty remarkable. But this paled into insignificance when, in 2020, Godfried Schreur, Sergio Mayordomo and I analysed in depth a redstart recorded by Godfried in Badajoz (Spain).[2] In one hour of song, this particular champion imitated no fewer than 51 other species, including such a motley crew as booted eagle, golden oriole (call and song), heron, nightjar, nightingale, Orphean warbler, willow warbler, partridge, green sandpiper, greenshank, little ringed plover (these three waders, incredibly, packed into the same one-second phrase), swift, bee-eater, grasshopper warbler... It proved to be a mimicry game-changer similar to David Attenborough's seminal recordings of the chainsaw-imitating lyre bird. But it need not be so exotic or far afield. Checking out this remarkable ability among songbirds is something any of us can do any day with a local starling or song thrush. You can indeed have great fun recognising in the chimneypot starling the greenshank or partridge, ghosted in from lake or field, or the swift brought down from the heights; the electronic squeal made by some cars upon being opened or closed with the remote key; or the ringtone of the neighbour's smartphone.

Poets and musicians over the centuries, without delving too much into the science behind birdsong, have certainly recognised and celebrated its mindboggling complexity and beauty. Witness Keats, Shelley, Wordsworth, Heine, Rilke, Beethoven, Schubert, Mahler, Messiaen to name only a few of an endless list. Well... endless... The sad thing is that today it has just about ground to a halt. Few artists today celebrate the miracle of birdsong and hardly any of us notice the blackbirds, blackcaps, garden warblers and robins performing these miracles in our own gardens. For we don't need to go far, that's the thing. From March to July gardens and parks, even towns and cities, in any part of the UK are all resounding with this beauty. We can, as it were, plug ourselves in at will, to this millennial evolution. As Keats says in his 'Ode to a Nightingale',

> The voice I hear this passing night was heard
> In ancient days by emperor and clown

We could even go back to that primordial musician with his hollow reed or bone. In that sense the song is immortal and part of the elegiac delight of listening to it is perhaps an awareness of our own short mortality against this timeless beauty.

Timeless, but certainly not invulnerable to time. Birdsong, nowadays, faces many threats—as many as those that hover over the birds themselves—like habitat destruction, pollution, hunting, to name only some of the worst. But probably the most direct threat to birdsong itself is... *noise*. Airports, motorways, industry all make it harder for us to hear the birds and harder for them to hear themselves. Many studies have shown that urban noise is changing birdsong at a remarkably and worryingly fast rate.[3] The nightingales of Berlin's parks have learned to sing louder to overcome traffic noise. In fact the decibels they now churn out make them technically *illegal* according to municipal byelaws. Great tits have raised their pitch instead of their volume, even though, intrinsically, lower pitches are more pleasing to the females.[4] This is already leading to the development of two different tribes within the species: one with the traditional tried-and-tested lower-pitched song and another with the noise-surmounting higher-pitched song; if a bird from one of the tribes wanders into the environment of the other, it will find it much harder to attract a female. Birds around major airports have had to advance their dawn chorus by up to half an hour to make themselves heard.[5] All this confusion is having an untold effect on the species' development, so steady for millions of years and now so sorely disrupted in just a few decades.

What can we do about this? Well, one simple thing is not to add to these noise or pollution levels. I am sure all birders, including twitchers, must want birds and birdsong to thrive, but we are not exactly setting a conscientious, conservation-minded example by careering off in a car around the country to tick off the latest egregious rarity found hundreds of miles from our homes. We have all done it to some extent because the lure of the new is real and appealing. In the final analysis, however, is that desert wheatear in Scunthorpe really more remarkable than the virtuosos performing unheralded in our gardens, parks and local woods, fields or heaths within walking or cycling distance? Maybe it is partly the collective loss of this ability to appreciate birdsong in our local patch that has driven us to seek overachieving thrills so far away. And in doing so we waste so much valuable time sitting inside the sterile bubble of a car or stuck in traffic jams, often forfeiting precisely the best part of the day too, dawn, when all the birds are vying to outperform each other. If we leave the house on foot or bike we do not miss one second of this tremendous free-to-air daily concert from the start to the finish of our outing. Keats wrote his ode while listening to a nightingale on Hampstead Heath, so things have certainly changed. Even so, from London we can still reach nightingales in a couple of hours on a bike (I have done it many times), while blackbirds, wrens, robins, song thrushes or chaffinches, all remarkable performers in their own right, can be heard in practically any part of the UK just by opening the window.

Simply that. Imagine if a stunning Vermeer painting were to be left hanging from our porches for four months every year and we had to beg people to look at it? That is what I always feel when pleading the case of birdsong. It seems to be counterintuitive. Such beauty should its own advocate be. Unfortunately, that is no longer the case. But, like that stubborn Shetlands diner, are we really going to keep our backs turned to this daily marvel? I have mentioned that elegiac tinge of birdsong, confronting us with our own mortality. There can also be a pleasing dash of altruism as we help to pass birdsong on unharmed and unchanged into the future.

17

The Sound of Summer

Arjun Dutta

I am very much a summer person. Summer means long, warm days and happy memories. And, of course, summer also means swifts. Whatever I do and wherever I go in the summer months, I know I will see and, just as importantly, hear swifts. While this is the case every year, 2020's arrival of swifts meant more than ever before. When the Coronavirus lockdown was announced in March, birds and the natural world assumed a greater importance in helping the nation to endure difficult times. A month doing very little except birding at home, away from my patch of Beddington Farmlands in South London, ended on 29 April when I returned there to see my first swifts of the year, flying over the lakes in the rain. Seeing more than 300 birds, all of which had most likely just arrived for the summer, feeding together in stormy weather was truly a spectacle to behold. The deep breath I took at that instant… the relieved smile… I will never stop appreciating that once-a-year moment.

With lockdown keeping everyone at home or within proximity of home throughout the spring and early summer,

I set myself a simple, little target round the start of May, and that was to make sure that, at least once a day, I spent at least a few minutes standing in the garden watching and listening to swifts. This was a task I pursued through the best and the worst times of summer, dedicating a little time every day to a bird that kept me going. Sometimes I took my camera out with me (I had always wanted to get some half decent pictures of them) and, although not amazing, I was pleased with the year's efforts, especially as room for improvement was always possible. Most of the time, however, I took just my binoculars, and simply stood there, watching screaming family groups fill the skies of South London.

The thing I love most of all about these birds is the sound—in fact, 'screaming' swifts have been subjects of interest for many centuries, for folklore and to general science. I find this screaming almost enchanting at times—peaceful, soothing and re-energising. After becoming obsessed with sound recording from early April, I eventually started to go outside with my sound recorder and mic on a mission to record them, which proved harder than I thought it would be due to the volume of other birds, traffic and various mechanical noises. Over the course of July, once young had departed the eaves of the houses in which they had been born, huge flocks of birds were visible and audible high above the houses, loudest at dusk. As a result, I found myself recording on most warm evenings until they left, once and for all, in the first week of August—these recordings are perhaps the most valuable on my xeno-canto account online.[1] Swifts really are the indisputable, enigmatic sound of summer.

As well as watching the aerial assassins from home, the second place I normally pay close attention to them is at Cheam CC, my cricket club, where I have played since I was seven. For years, swifts were audible every night at training or during matches, with every bird bringing an added moment of happiness. Thankfully, matches could take place again in 2021;

for much of 2020 though, the pandemic's restrictions prevented me from visiting Cheam at all. Instead, I had to focus on the swifts at Beddington. Though brilliant for patch birding during lockdown on warm, sunny days, when the reedbeds were alive with the sound of warblers, the meadows buzzed with insect life, the lake held several little ringed plovers, and the scrub bore rattling lesser whitethroats, I started to go to the site for a specific reason—a purposefully planned method of birding I like to describe as 'twitching the weather'. Twitching is typically associated with travelling specifically to see particular or rare birds. Twitching the weather at Beddington was instead an occasion to see one of my favourite birding spectacles. When storms are predicted, with heavy skies of wind, rain and cloud, hundreds of swifts from the local area all gather in a feeding frenzy of more than 400 birds to pick off insects over the lakes. On these days I have noticed that swifts are nowhere to be seen at home, which means that birding from the garden is much less enjoyable; going to Beddington on these days, 11 times in all across that summer, enabled me to get within metres of swifts and swallows as well as hobbies, well known predators of such birds.

In addition to what can be seen simply by watching these special birds, they are also fascinating to learn about:

- They mate, roost and sleep on the wing—they are thought to remain in flight all winter in Africa, never landing until the next breeding season. Juvenile birds often do not land for several years, until they breed!
- The routes they take to reach Africa are only theoretical—there is still so much more to learn about them.
- Their flying speeds are incredible, even when migrating. One is known to have travelled nearly 1,500 kilometres in 3 days.
- They can fly at altitudes of up to 6,000 metres.
- Some birds may fly over 6.5 million kilometres in their lives—equivalent to eight trips to the moon and back.

As mentioned above, there is still much to learn about swifts. Their migration routes, wintering grounds, flight patterns and behaviour are all fascinating subjects yet relatively unknown compared with many other species. The work currently being undertaken by the British Trust for Ornithology is particularly interesting. Researchers are equipping swifts with geolocators, small devices which provide information about where these birds spend their time. The insights generated by this research about little-known aspects of their lives, particularly during their migration and in their wintering grounds, may help ensure their long-term survival. Few birds are as incredible as the swift!

The UK is home to over 50,000 pairs of common swift, alongside many more non-breeding individuals. However, across the globe, there are at least 113 described species in the Apodidae family, which includes swifts, swiftlets, needletails and spinetails, only eight of which have been seen here. Yet, when a rarer swift is sighted in the UK, the excitement felt by many birders always seems to be that much greater than the excitement caused by other rare birds. For instance, the discovery of a white-throated needletail on Harris in Scotland remains a memorable rarity from the past decade of British birding, despite the fact that it was tragically killed by a wind turbine. In recent years, sightings of a white-rumped swift, numerous Pacific swifts and a showy little swift all spring to mind in further demonstrating a love of the swift family shared by thousands of birders and nature enthusiasts. Going abroad to Europe has allowed me to focus on some of the species of swift less often seen back home—watching the brutish elegance of Alpine swifts as they swirled effortlessly above Athens while on a school trip is a memory I look back on with fondness. At certain times of year, Alpine and pallid swifts are prized 'scarcity' finds for many dedicated birders in the UK. They really are a family which is almost universally enjoyed by everyone—from twitchers and foreign travellers, to photographers and sound recorders, as well as casual and patch birders.

By the end of July, the ever-approaching, inevitable departure means each encounter is all the more valuable to me than the last. I have not enjoyed school in recent years as much as I would have liked. With it being my final year, I was less enthusiastic about returning in 2020 than I would have been in a normal year. However, seeing my final swift of the year on my first day back to school on 2 September was possibly the most perfect and reassuring way to return that I could have asked for. All through the winter I always long for their return, for the feeling these miniature birding missiles inspire in me. Year on year, they never fail to do wonders for my mental health and frame of mind. I know they will continue to do so, which is why they are so meaningful. Sadly, like so many birds, their population is rapidly declining—one source suggests by as much as 53% since 1995. Swift conservation remains crucial in sustaining populations. While the future can often feel bleak and depressing due to the biodiversity and climate crises facing the world, there is so much that can still be done to help birds like swifts. One day, my dream project would be to set up a school or university scheme to install swift nest boxes on buildings. One day…

With all this in mind, it is perhaps necessary to reflect on why a single bird species can be so significant in the context of the environment and its inextricable link with people. As I hope I have demonstrated, travelling large distances to experience meaningful or enjoyable wildlife encounters is not a necessity. Nor is having any specific birding equipment. In the past year, my watching and listening to birds like swifts has largely been completed on foot, either with a quick walk to the garden or to Beddington Farms. Cycling was also a viable option. Yet most of my happiest moments of swift watching have taken place in my garden, or even from my bedroom window. My garden list of 81 bird species—105 including those which have flown over at night—emphasises how, even though I live in South London, using cars and higher-carbon modes of transport are not needed to appreciate the natural world, or benefit from the joys it can bring. A winter's walk simply to watch stonechats can provide greater therapy than anything else to which one might otherwise turn. Appreciating nature does not have to mean becoming a full-time 'birder'—it can be as simple as spending more time outdoors. Hopefully, the future will be brighter for both nature and people, and more of us, of all ethnicities, genders, ages and socioeconomic backgrounds, will be able to benefit from the peace and joy brought by birds and wildlife!

18

Birding in the Yorkshire Dales

Steven Ward

Birding biographies can have their ups and downs. While I had once been a budding Young Ornithologists Club member, I stopped birdwatching for 13 years, only taking it up again at age 23. When I was a child the obvious place to go birding was the short hop through the backfields, on foot, down to the River Ure, by the town of Hawes in the Yorkshire Dales, where I live. But once I started to enjoy birds again in my twenties, I spent more than a decade incorporating not a small amount of motorised travel into my birdwatching, reaching a head in 2017 when I attempted a year list, travelling fairly extensively over Northern England and Scotland by car, with the odd twitch thrown in. Despite going to some amazing places and seeing some great birds, instead of finding it fulfilling, I felt frustrated. It dawned on me that I seemed to be charging around in a car almost as much as I was actually birding. It ultimately felt like a race—an unnecessary race against only myself, at that.

Fast forward to the present day, and I am now part of a growing number of birders embracing and enthusing about a local patch and low-carbon birdwatching with travelling undertaken on foot, by bicycle or by public transport. In my case, and I am sure for many others, this new focus was helped by the Covid-19 lockdowns, during which driving for recreation was prohibited.

The exercise benefits of being a green birder aside, I have only recently discovered the real delights of garden birding or, to be more specific, birding from the garden. I had seen on social media a growing movement of people doing this, and so it was not the most original of ideas, and I felt a little late to the party. But, having a little more time on my hands after having been furloughed during the first lockdown, I was keen to get involved in this virtual get-together, and so I took to watching garden birds and staring at the horizon.

Most days, around lunchtime, after getting the most junior member of the family down for his midday nap, my five-year-old and I would decamp into the garden and driveway with camping chairs and notebook, optics in hand, to gaze skyward. I consider myself lucky to have a splendid panorama of fells and valleys to look out upon, with the view extending to hilltops almost six kilometres away.

On these largely warm and balmy April and May days, we hit garden birding heights I would scarcely have believed possible when starting out. There were modest highlights such as the first-ever seen-from-garden stock doves, scoped at a couple of kilometres away, flushed by buzzards on a fell-top scar which, with my now more prolonged garden observations, became near daily flyovers throughout April. Only my second seen-from-garden green woodpecker was another goodie, yaffling away from a wooded garden several fields away. My first seen-from-garden blackcap record was a male singing from the garden area of one of the town carparks, now deathly quiet with the lack of human visitors. There was a distant, but daily, male redstart to enjoy when he finally turned up on 23 April in the aforementioned yaffle's trees, which could be scoped from the garden.

As the month progressed, more exciting and exotic records included several days' worth of peregrines. One memorable session in late April had two immature birds up in the air, flying low over our small market town, causing much consternation in the local rookeries. What menace they exuded! Given that peregrines are much scarcer than they should be in the Dales, largely due to persecution by humans, these were a real thrill to see. Further excitement came from several red kite and osprey sightings. At least one of the latter seemingly toured Upper Wensleydale throughout the spring and summer. With my wife and child I even managed to follow up a garden sighting with a sprint to the top of the next-door field to see the bird dive and catch a fish from the River Ure in the valley bottom.

These sightings, however, pale into insignificance compared with an event on 18 April. While scanning the horizon, I noticed a very distant group of birds, seemingly flying in my direction. From these initial views, I thought they were most likely starlings. Managing to keep the scope steady, I persevered with the flock and as they drew closer, details beginning to emerge, their identity eventually became apparent. Thrush family. Check. White chest crescent. Check. They were ring ouzels! Thirteen of them flew by, around 200 metres away. They breed in the nearby hills, but never would I consider getting them on my garden list. As they disappeared, strangely flying in a southerly direction, my heart was pounding and, even more noticeably, my legs were like jelly. Never before or since have my hands fumbled so much to send out a Tweet. Such was my obsession with increasing my lockdown garden list, I also took to scoping spots where in the past I had seen red and black grouse, short-eared owl and hen harrier. No luck so far, but I remain hopeful for the future.

Of course, sky-watching from the garden is not all about picking up flyover records of unexpected or rarer species. Watching the skies regularly from a fixed point gives you the chance to spot subtle patterns in movements or behaviour which you would not have noticed otherwise. The almost constant passage of lesser black-backed gulls throughout late spring and summer is a great example. Virtually all adults, are these simply wandering non-breeders? Or perhaps bona fide breeding birds, maybe from inland breeding colonies such as in the Forest of Bowland, travelling afar on feeding forays? There are always more questions than answers.

Regular visitors and breeders in the garden are also a source of questions, especially when you pay attention to the behaviour of individual birds. During quiet passage periods, I discovered breeding attempts by jackdaw, blackbird, song thrush, robin, dunnock, pied wagtail and house sparrow in my own and adjacent gardens. It is these regular day-to-day observations which can keep the interest piqued.

I am also a keen cyclist and I have found that it is possible to see, hear and record good birds while biking. As virtually all of my rides are from the front door, my riding largely takes place in north-western North Yorkshire and south-eastern Cumbria. In the summer of 2020, after the first Covid-19 lockdown, I would head up onto the fell-top bridleways a couple of evenings a week. This year must have been a 'vole year' as almost every ride out garnered sightings of the magnificent short-eared owl. Later in the season, away from the more heavily managed moors at least, breeding pairs appeared to be having a bumper year. As June turned into July, fledged broods were regularly seen, the juveniles flapping off the top of drystone walls just ahead of me, with the family party curiously eyeing me from above, one evening disappearing and reappearing eerily in and out of the mist. I am very keen on my sightings having a real scientific purpose, and so I submit my records on the BirdTrack website of the British Trust for Ornithology (BTO), by which means they can reach the Rare Breeding Birds Panel, helping them keep track of these rare breeding owls.

Although redstarts are not uncommon summer visitors to the Yorkshire Dales, their Atlantic oakwood cousins, the pied flycatcher and especially the wood warbler, can be hard to catch. Last year, 2020, I picked up my only pied flycatcher of the year on a mountain bike ride in a small wood from which they seemed to have disappeared since the last BTO Bird Atlas period. Had I been journeying by car, I would have completely missed the sweet-sounding, jaunty ditty. It held territory throughout the season, and, I would hope, bred successfully, possibly spilling forth some humbug chicks to add to the now scarce Upper Wensleydale population.

Even more satisfying, on a necessary lockdown trip to the supermarket by car, I briefly paused at a likely stop for a chance of finding a garden warbler. To my surprise and delight, I instead encountered a singing male wood warbler. A few days later, I incorporated this area into a daily exercise mountain bike ride to get a 'green' wood warbler onto my lockdown list.

Although I have now become very keen on my green, non-motorised bird lists, it has to be said that due to the remoteness of the uplands and the poor public transport network, everyday life is highly dependent on cars. Though I have pretty much stopped the modest amount of twitching that I did, and have substantially cut back on past monthly 'big days away' involving many road miles, I do still use the car in small doses for local birding. I also appreciate that not everybody can give up driving altogether if they are to continue to be able to go birding, often due to their age or health problems. Particularly in the uplands, decisions about how we enjoy and study birds are not always straightforward, especially when access to certain areas for collecting data in poorly monitored areas may require some driving.

Likewise, showing the kids our special Dales' birds requires the car. Very small legs and very large hills would not make for a serene experience, so the car gets a quick run out on these occasions.

For the sake of our birds and the planet, I would hope the low-carbon birding movement encourages birders to reassess the way in which we all do our birdwatching. Whether it be a walk around our local patch from home or a trip around the continent by train, we can all do our bit to keep carbon emissions down. No one can be the perfect low-carbon birder all of the time, but let's play our part in making the planet a safer place for ourselves and the birds!

19

TG42

Tim Allwood

My journey into low-carbon birding has fundamentally changed my life for the better. It began with a family holiday to north Norfolk in the 1980s, which sowed the seeds of what has become a lifetime's addiction. My parents allowed me to wander the reserve at Cley alone and I remember being mesmerised by the number of waders beetling around on the scrapes. The presence of a Wilson's phalarope was an introduction to the excitement and fascination engendered by birds from far-flung places, and I wondered if one day, I too, would find myself settling in a flint cottage in a village on the Norfolk coast.

After leaving university and spells working for the Royal Society for the Protection of Birds, teaching overseas and travelling to see birds, I found myself teaching in Norfolk. This was the point when my life really began to change. Teaching climate change to schoolchildren was beginning to make me question my approach to birding—and life in general—as I developed an understanding of the mechanics and maths behind the science, and what climate change meant for our future.

Being based in Norwich, I was close to numerous well-known birding locations and many of my weekends were taken up driving to the coast for a few hours to lose myself looking for birds in some of the quieter spots I had come across. The frustration at only being able to do this by car and at weekends grew and eventually I took the plunge, buying a house in an east coast village called Sea Palling. It was not a flint cottage, nor was it in north Norfolk, but it was the best decision I have ever made!

Other major lifestyle changes followed: I decided in April 2007 that I had taken my last flight. I could not stand in front of children and describe the damage we were causing to the climate system and expect them to take me seriously if I was not acting like it was the major problem I was telling them it was. Taking on an allotment and growing a lot of our food came next, along with purchasing a good quality bike for birding.

My obsession is my adopted birding patch of TG42. This is a ten kilometre square, only half of which is land, the coast bisecting it from the north-west corner to the south-east corner. It is essentially rural, dominated by arable crops, rough pasture, marshland, broads and meres, with extensive coastal scrub and cover. You could not ask for a better combination of habitat types to attract a wide selection of bird species, all year round.

Initially, my adventures here were all about the birds. I spent long days searching for scarcer migrants and, as I began to meet people in different places across the square, I made new friends and acquaintances, with my birding walks and rides taking on an unexpected social aspect. It was not long before I met Andy Kane, a fantastic bird finder. A couple of rarer birds in my first autumn (Radde's and greenish warblers) helped break the ice and we became good friends, destined to share many fantastic birding moments. By the end of my first year, in the village pub or around the local area, I would be asked about unusual birds that people had seen, or what special birds were around. Striking up friendships with locals had the added bonus of gaining me access to private areas in the square. I found that politeness and patience work wonders, and before long I was able to bird in a couple of fantastic large gardens, on farmland away from roads, and on the broads by boat with Andy. Being able to bird such places in peace and quiet with the real possibility of rare birds just around the corner was a dream come true. I particularly remember evenings heading out on the boat to Rush Hill and Swim Coots scrapes, wondering how I had been so fortunate to end up there.

Living on your patch means that birds simply become a part of your daily life. All the time. Whenever I leave the house, birds are always high in my mind, if not right at the front. A glaucous gull on the way to the shop, tundra bean and Ross's geese while digging the allotment, my greyhound plucking a bittern out of pathside vegetation (both unharmed, the dog much more scared than the bittern), a wheatear on a rooftop during the short walk to the pub, a flock of bee-eaters while putting in cabbages, a yellow-browed warbler as I pulled onto the drive, or even a stone curlew while lying in bed in the dead of night—all chance encounters while just going about my daily routine. Even the journey to work can be a delight—especially passing flocks of geese and cranes in winter.

TG42 is particularly good for both migrants and vagrants at almost any time of year. Seeing regular migrants on a short walk from your door just cannot be beaten, and every day has a potential surprise in store. Over the years, we have been very lucky with Radde's, dusky, Bonelli's, subalpine, Hume's and arctic warblers, red-flanked bluetails, eastern yellow wagtail, tawny pipit, a Pallas's grasshopper warbler that we spent two hours nearly treading on, black lark, pied and desert wheatears, and lesser grey and isabelline shrikes among others. Hickling and associated wetland areas have produced stilts;

Baird's, white-rumped, buff-breasted, broad-billed and marsh sandpipers; Kentish plovers; Pacific and American golden plovers; long-billed dowitcher; collared pratincole; squacco heron; Franklin's and Bonaparte's gulls; and Caspian and white-winged black terns.

The sea at the bottom of my road has been a major part of my birding and I soon became captivated by its unpredictability—both in terms of the birds and the weather. A day up at Andy's home in our foxhole, behind the windbreak, and sitting under the umbrella with the wind howling and the skuas screaming by, is something to experience. I never get tired of it. Over the years, effort has been repaid with sightings of Fea's petrel, Cory's and great shearwaters, white-billed and black-throated divers, long-tailed skuas (probably my favourite species), king eider, Sabine's gull, and rarer grebes. Watching the movements of wildfowl, waders and terns can be as exciting as seeing rarer species—if not more so—as the excitement and interest are prolonged. I remember one day with nearly 1,000 Manx shearwaters, a crazy day with nearly 6,000 gannets streaming past against an angry black sea, and another with almost 2,000 red-throated divers in mini squadrons, coming north out of the sun in the early morning.

The sound of large flocks of geese over the house on winter mornings, or the sight of birds feeding in frosty fields is a wonderful thing to experience. I am seldom happier than when searching through 'pinkfeet' for rarer geese, such as tundra and taiga bean, brent, Ross's and snow, Greenland whitefront and wild greylag.

Now, nearly 15 years after moving here, driving more than a few miles to see a bird has become anathema, and I will take my trusty charge-plug cycle instead. Despite having notionally limited my horizons, I have found that I derive much more pleasure from a deeper understanding of my local square, and being part of its life, than from driving to see (or not see!) something on the other side of the county or country that I have no meaningful connection to. An understanding of the climate emergency we are in, and the politics of energy production, have resulted in low-carbon birding becoming an integral part of my life. I cannot teach children about climate change, or talk to other birders about it, if I am not behaving in a way that reflects the message I am conveying. Other lifestyle changes have seemed like a natural progression to me—giving up meat, markedly reducing my dairy intake, buying second-hand clothes and generally trying to have a lower environmental impact wherever possible. These changes are not as hard to make as people might think. The hardest step to take is the first one. Once you have made the decision to go low carbon, other things seem to follow naturally. Knowledge is power, and learning about the science and politics of climate change and resistance has strengthened my resolve to act in accordance with my words and set an example to others, showing that positive and life-enriching changes can be made. My pace of life has slowed and my birding has become much more enjoyable. The stresses and conflicts of driving long distances have been replaced with walks and cycle rides, the optimism of potential discoveries and the sense of wellbeing that a less harmful, healthier, lower-impact lifestyle gives.

In the time I have been here, as well as the climate emergency getting much worse, I have also noticed a clear decline in many species of bird. Skua numbers have decreased markedly, wheatears are well on the way to becoming scarce, yellow wagtails are now a surprise on a spring morning rather than an expected feature, and wader and duck passage on the sea is certainly not what it was. This has made my experiences somewhat bittersweet as, despite the joy I have experienced, I can see things are deteriorating and I know we are entering difficult times. I worry about the future for my daughter and the children I teach if we continue to ignore science's warnings on the climate. I feel some guilt at having been part of the generation responsible for some of the damage done to the planet and this fuels my passion and resolve to do what I can to make amends, however small and insignificant my actions may appear on their own. It is simply the right thing to do.

Of course, despite the abovementioned issues, it is easier to be content with your birding if you live somewhere that still has such rich possibilities as I have described, and I fully understand that. Living on my patch and becoming a part of it, birding locally on foot or by cycle, engaging with my local community, forging new friendships, adopting low-carbon birding and joining the political fight against climate change, have all enriched my life hugely. Along with my teaching, these things have given me direction and purpose in an often confusing and dispiriting world. If you can make such changes, I would highly recommend you do so. You will not regret it!

20

Shrikes from the Bike

Dave Langlois

When I started birding back in 1960 red-backed shrikes still bred in London. Fast forward only 28 years and I am watching Britain's last breeding pair in Santon Downham, Suffolk. The experience of witnessing this bird's fulminating extinction during my formative years (there have been recent, sporadic breeding attempts in various sites in Devon or Shetland) has added a sheen of subjective fascination to its many intrinsic merits: its pocket-battleship build; highwayman stance; magnificent livery; hunting prowess, with gliding pounces and dashing flight; its secretive almost hieratic lifestyle; and its astonishing, Mediterranean-lapping migration route. There is no other bird like it.

Although my family and I live in La Vera, Extremadura, for most of the year, in 2011 we started to spend the summer months in Asturias in the north of Spain. On my rides in this wonderful cycling area, initially at random, I started to look for shrikes breeding in the surrounding countryside. With increasing effort and targeted dedication over the ten summers since, I have managed to find a total of 260 territories. The study/cycling area is a rough rectangle along the Asturian coast marked out by Villaviciosa in the north-west, Infiesto in the south-west, Llanes in the north-east and Arenas de Cabrales in the south-east, adding up to a total area of around 1,800 square kilometres. The total kilometres ridden every summer in the three months of surveying from late May to mid-August add up to about 3,000.

Cycling is the ideal way to census these birds because you can cover a lot of ground and pick them up at any point by eye or ear, stopping at will. The male shrike in particular is often very eye-catching when perched atop a bush or on a cable, but they can also vanish within their large (sometimes shared) territories for hours at a time. And they can hardly be said to sing at all. Not only do they seldom perform, but the song itself—a typical shrike melange of mimicry, squeaks and twitters, some of it quite sweet—is inaudible at more than a few metres. I have heard it most (least rarely) late in the season just before they migrate, but this might well be because it is only when other birds have stopped singing that you can hear it. It is partly this customary silence that makes them so hieratic; they just suddenly appear, noiselessly, perched atop a fence post or bush you have checked in vain scores of times before.

The most useful calls for picking them up from the saddle are, firstly, a nasal territorial call given out by the male, above all in the early part of the season, and, secondly, a young-escorting alarm given out by both parents (and the elder young) once the brood leaves the nest. The former, often produced from the top of a tree, is rather like a flat, inflectionless sparrow chirp that has been slowed down and played backwards. Sometimes it is distorted into a quirky squeak, somewhat like the nightjar's flight call, or one of the carrion crow's stranger offerings. That is about all you hear for two months or so. Then the young appear, from late June onwards, and the adults start to make their *shi-shik* (meaning its name might be onomatopoeic?) call, sounding like a bike's two wheels passing over the same dry leaf. They may also string them out into a longer sequence of *shi-shi-shi-shik*, but always with a horizontal emphasis rather than the vertical staccato *chaks* of the *Sylvia* species and the stonechat. The young, sometimes from the nest but above all as food-begging fledglings, produce an unmistakable, high-pitched and very far-carrying banshee squealing. This makes July and early August fairly hectic: you have to cover as many territories as possible while the birds, young and parents, are most vocal. Once you get your ear in, you can easily pick up all these calls from a moving bike. Even 71-year-old ears that have lost all the highest frequencies…

The bike I use is a goodish carbon fibre road bike. This lends itself to much longer daily journeys than a heavier, higher-friction mountain bike. The drawback is that it limits you to paved roads, but eastern Asturias has a pretty good network of minor roads. I also sometimes hide my bike in bushes and explore a track on foot or even carry my bike cyclocross style along a promising dirt track linking two paved roads.

The 260 territories break down into 71 individual territories and 67 lax colonies of two to ten pairs, making 138 sites in all. In 2021, 194 of these 260 territories were still active. This figure hides a positive skew: the new territories found in 2021 (33) were, by definition, occupied. Discounting these new territories 161 of the 227 (70%) territories found up to the previous year were occupied. This same percentage (occupancy of sites found up to the previous year) has fluctuated between 50% and 70% from 2015 to 2021 with no clear upward or downward trend. Future years might clarify the situation. The survey, of course, would have been more scientific (but far less enjoyable) if it had been restricted to a sample of the same 100 territories each year. But this has been more a structured leisure pursuit than a serious scientific study. As for productivity, in 2021 young were seen in 63 of the 138 sites. Again, a caveat: this figure is almost certainly an underestimate; it is impossible to cover all the sites adequately in the month or so when the young are on the wing and vocal.

Typical shrike habitat in this part of Asturias consists of livestock pastures with hedgerows, scattered bushes and trees, and sometimes woodland edges with easy access to such open ground for hunting. One of the most surprising findings of this survey is that they tend to eschew ground above 350 metres elevation even though this habitat extends much higher. The highest territory found in this ten year period is the mountain pass Moande, at 607 metres, where they bred and produced young until 2014 and have been absent from ever since. The highest extant territory, another outlier, lies at 452 metres in Villaverde, Amieva. In general the territories break down as follows: 39% lie between elevations of 0 to 100 metres; 35% from 100 to 200 metres; 20% from 200 to 300 metres; and only 6% at over 300 metres. Why should this be? In other parts of Asturias they breed at elevations of well over 1,000 metres. It is hard to imagine what the answer might be, and it would be very interesting to find out why with a more in-depth, dedicated study.

Lastly, a few words in favour of this way of birding in general. All I have said above about the shrike applies to other species too, even more so as others are more vocal: you can pick them up splendidly as you pedal along. In Asturias this might be marsh tit, yellowhammer or black woodpecker on the lower part of the climbs; wheatear, snow finch, Alpine accentor or wallcreeper (with luck) at the top; golden oriole, song thrush, blackcap and Iberian chiffchaff in the woods; redstart, blackbird, swift and swallow in the villages; buzzard, honey buzzard, hobby, short-toed eagle or Egyptian vulture in the sky; garden warbler, grasshopper warbler and tree pipit in the scrub; and raven, peregrine, chough and passage shearwaters along the coast. There is always the occasional, delightful surprise. In June 2021 I stopped to look at what I thought was another shrike atop a tree and found I had trained my binoculars instead on a summer-plumaged rosy starling.

Passing by in a car I would never have noticed it and could not have stopped on that stretch of the road even if I did.

The combined sense of athletic achievement and aesthetic fulfilment is a heady mix, none headier, enhanced by the altruistic satisfaction of adding no noise or pollution to this punch-drunk world of ours. In my experience, birding by bike is a solitary pursuit; after all, few birders want to cycle 100 kilometres and even fewer cyclists want to stop for hours listening to a redstart running through its roster of imitations or watching a family of shrikes hunting in a sunlit meadow. But, hey, you might be lucky enough to find like-minded and like-legged companions as, hopefully, our numbers grow in the future.

I particularly recommend starting spring and summer rides before dawn, something even fewer people do. It is a joy. Every climb at that time of day is a triple crescendo of sound, height and light, as the dawn chorus takes off, the hilltop comes closer and the day waxes. The beguilement of the birdsong even eases the hurt... Every stunning viewpoint is your own exclusive property instead of a crush of cars and camper vans. The rising sun peeps over the mountain top to flood you with warmth, opening your soul like a butterfly's wings, while the valleys are drowned in a sea of mist below your feet. You feel... well... on top of the world.

21

The Best Kind of Golden Oriole

Gavin Haig

My very first bird book was the 1966 edition of *British Birds* by F. B. Kirkman and F. C. R. Jourdain, a gift from grandparents who had spotted in the nine-year-old me a fledgling interest, and responded perfectly. Although long gone now, many of the book's plates live on in my memory. The great grey shrike, for example, perched like a sentinel on its shiny page, the bird's outline deeply indented by my boyish efforts with tracing paper and pencil. But one species shone like no other. I can see it vividly in my mind's eye: yellow as a lemon, with black trimmings and a blood-red bill. The golden oriole was simply too gaudy to be possible, and I could never imagine seeing one in real life.

I might have lusted over its picture, but here was clearly a bird meant only to decorate books, and quite unattainable.

Most of my boyhood was spent pursuing coarse fish with rod and line, and birds were a side interest only. At least, they were until the spring of 1978, when an enthusiastic fellow student introduced me to 'proper' birdwatching. Suddenly, impossible birds were flying off the page and into my life. I was hooked. These feathered wonders materialised right there in the local countryside, or during field trips with the college ornithological society. I learned what it meant to 'tick' a new bird, and the need to keep a 'list'. And then one day I was introduced to 'twitching'...

Dave Collins—generally acknowledged to be the sharpest birdwatcher on campus—had found a rare bird at the nearby Wraysbury Gravel Pits. The following day, my friend and I cycled over to try to find it. We scoured the area near Wraysbury Station, where Dave had heard it singing from a wooded island, but there was no sign. Looking back I cannot help a wry smile. The bird which had sparked this hopeful endeavour—my first ever twitch—was the very one which had long ago seemed so out of reach: a golden oriole.

By the early 1980s I had left the fishing behind. Birds were all-consuming now, but I was no longer just a birdwatcher. I had become a birder, like all my friends. This was not just semantics. For me, the word 'birder' had a more dynamic, thrusting connotation than 'birdwatcher' and, as the decade progressed, became synonymous with an increasingly potent culture centred upon rarity. Birders were not really out-and-out twitchers, but most would drop everything for a new bird, a tick on the list. And they were more than ready to travel for their birds. I certainly was. Living in suburban West London, I drove to the Staines area and Colne Valley for so-called local birding; to East Anglia, Kent, Dorset and almost anywhere else close enough for a day trip. And of course there was frequent twitching.

It was probably the spring of 1983 when my friends and I learned about the secret site, and very quickly a trip was on. Arriving at first light one May morning, we were barely away from the car and could already hear our quarry. The haunting, fluty whistles truly belonged in a rainforest, not this East Anglian poplar stand. And, 15 years after making me aware of its improbable existence through the glossy pages of a book, the golden oriole finally came to life. That first male has got to be one of the least anti-climactic birds I have ever clapped eyes on: visually, aurally, in every way stunning. I was completely blown away. More prosaically, my burgeoning life list increased by one, and I expect we spent the rest of the day thrashing round Norfolk from hotspot to hotspot, like we usually did back then.

As the 1980s segued into the 1990s something happened to my relationship with birding. Something bad. As that latter decade progressed I found myself twitching less and less. Adding ticks to lists was becoming an increasingly empty pursuit. Something fundamental was missing. It was as if I had fallen out of love with my hobby.

Looking back at that period, there *were* times when I enjoyed my birding, but almost exclusively they involved a holiday. The Isles of Scilly featured, also Norfolk. No twitching, just pottering about, treating the holiday venue like a local patch and being happy just to see what turned up. In May 1996 my wife and I were on Scilly with a non-birding friend and her son. Strolling along a lane together, our friend suddenly exclaimed, 'Oh! What's that bird?!' and pointed at a vision in yellow and black perched on a low fence. It was a stonking male golden oriole, which kindly obliged us with amazing views for a memorable minute or two.

But still—holidays excepted—birding and I had now become like awkward strangers. Thankfully, though, that is not the case today. My renaissance began with a move to East Devon in late 2002, and continues now in West Dorset. In a minute I shall explain how I fell in love again. But first, a digression.

In 2005 I began to write about my birding. Initially I contributed to an online forum, but in 2008 I began a personal blog. Most blog posts recount my everyday birding exploits, but not all. Among other things I do write the occasional opinion piece, and it was one of those which led indirectly to the words on this page.

With Coronavirus a sinister, looming threat, February 2020 was a time of worrying unease. As a kind of personal antidote I wrote what I thought was a light and cheery post entitled 'The Twitching Thing'. Though no longer a participant myself, I discussed why twitching had once appealed to me, and concluded: 'Why knock it? If there is one thing all of us need in this world, it is a bit of light relief...' Among the reactions to that blog post were two unexpected flies in the ointment. 'What about climate change?' they asked, like a couple of party-poopers complaining about the loudness of fun. Okay, so twitching is obviously a high-carbon activity, but, nevertheless, I was slightly miffed at this killjoy response, and ignored it. Crucially though, the point had hit home.

I should be frank here. Personally, I do not believe humans remotely capable of reigning in any of the various forms of galloping chaos we have inflicted upon this planet, climate change included. However, at the time of writing I am 62 years old, and appreciate that younger, cleverer minds think otherwise—or at least *hope* otherwise—and that I have a responsibility not to make things harder for them. The birding culture I grew up with, that matured and held sway in the 1980s and 1990s, and still thrives today, is clearly not compatible with a climate crisis.

In the weeks and months following 'The Twitching Thing', we experienced a nationwide lockdown due to the spread of Covid-19. Birding was suddenly limited to what you could see or hear from home, or during a daily exercise walk. And then, as the number of infections and fatalities moderated a little, and lockdown began to ease, news broke that several UK twitchers had apparently flouted Covid-19 travel restrictions in order to see a vagrant tern in the Republic of Ireland. That news was not received well by the wider birding community, and the actions of those involved robustly criticised. When my blog's blithely rose-tinted look at twitching attracted mild censure, I knew deep down that resentment was an immature response. The world is very rapidly changing for the worse, and any desire to preserve the birding status quo, to continue doing what we have always done, seems not only selfish and short-sighted, but deeply unfair to the younger generations. It sets such a poor example. For me that tern twitch crystallised the issue and I quickly found myself bashing away at the laptop. The resulting piece was entitled 'The Elephant in the Room'. In the birding community I do not have much of a voice, but that somewhat knee-jerk blog post rapidly garnered more views than anything I had written previously, and had evidently struck a chord. The 'elephant' is low-carbon birding, of course. And ignoring it should no longer be an acceptable option.

I do not feel I have experienced some kind of epiphany, but rather a slow realisation that the birding ethos I embraced in my younger years—which still pervades the modern scene—belongs in the past and should therefore end with my generation or sooner. Preferably sooner, because my generation is actually in a great position to set a good example.

I will never be a low-carbon birding zealot, but the very least I can do is share in this essay what it was that rekindled my love for birding, because doing so will hopefully shine a favourable light on a fairly low-carbon way to enjoy this hobby. I could sum things up by saying that this has involved adjusting my attitude. But that is a bit vague and, rather than waffle around trying to explain what I mean, I think it would be better to outline the three simple principles which basically underpin my approach:

- Stay local—Local is a relative term. I am fortunate to live three miles from the West Dorset coast. For me, a five or ten-minute drive is local. You might live even closer to some good birding, perhaps just a walk away. I can think of one or two city dwellers who have a very local birding area, but also periodically travel to a favourite regular spot further away for an improvement in scenery and prospects, a spot they then treat much as an extension of their patch. Any of these has got to be a step in the right direction away from belting all over the country every weekend.

- Have modest expectations—The modern British birding scene worships rarity like a god. There is nothing wrong with getting excited about a rare bird, but a constant diet of the things can lead to a warped view of reality and a jaded palate. Certainly I found that to be the case, and it became increasingly difficult to find satisfaction from local birding. The fact is, rare birds are rare, and scarce birds, scarce. Unless your local spot is very exceptional, that fact will dictate day-to-day proceedings. However, you soon come to appreciate that 'rare' and 'scarce' are also relative terms and—once your baseline has been recalibrated— that your local area is just as capable of giving you the rarity buzz as anywhere else. In the meantime, there will definitely be countless ways in which modest, everyday birds bring pleasure.

- Find your own birds—I cannot emphasise this one enough! Finding your own birds can be an immense source of satisfaction. Speaking personally, few things in birding give me greater pleasure. Let's face it: if you always travel to the same place as everyone else, at the same time as everyone else, the chance of finding your own birds is very, very small. On the other hand, if you regularly beaver away at a quiet local spot where few other birders go, the self-finding rewards will inevitably come. It is one thing to see a handsome male redstart at some popular, well-hammered location because a group of birders were already watching it, but quite another to have one unexpectedly pop out in front of you at your local patch. And nice, local patch finds will definitely happen. They will happen a lot. And will occasionally be on another level to redstart.

One of the most exciting moments in my recent years of local birding occurred in late May 2020. As it was a quiet time for migrants, I had just spent a few minutes photographing a yellowhammer. Starting along the path once more, I was stopped in my tracks by a thrush-sized bird flying from left to right in front of me. I knew what it was even before raising my binoculars, but the glorious, magnified view through lenses made me exclaim out loud: 'Oh, wow!' It was a golden oriole—a female or young male—a vivid mix of bright greenish-yellow and black. I watched it fly away into the distance, over a tall hedge and out of sight.

Relating this tale now, this fleeting encounter with a scarce bird, I cannot help but review more than half a century's acquaintance with the golden oriole. I have seen brighter individuals—beautiful, lustrous males— and enjoyed better, longer views too. But never have I seen one that meant so much. Because it was local, it was unexpected, and I found it for myself. And to my mind that is without question the *best* kind of golden oriole.

22

From Angst to Tranquillity

Jonathan Dean

In 2008—just a few weeks into our relationship—my partner mentioned to her octogenarian Portuguese grandmother that I liked birds. My partner's grandmother took my appreciation of birds and the natural world to mean that I had a generally calm and reflective demeanour. When this conversation was later relayed to me, I laughed with mocking disbelief. Like many other birders, my passion for birds was not, and at that time had never been, one of peace or tranquillity. Although I often justified my passion for birds on the basis of its supposedly positive effect on my mental health, in reality there was a strong element of self-delusion about this. My relationship with birding was, to a large extent, marked by stress and anxiety.

I started birding in 1991, aged nine, and my neurotic and compulsive tendencies were quick to emerge. The more birds I saw, the more anxious I became about the birds I had not yet seen. As a result, in my early teenage years I expended vast amounts of psychological energy trying to figure out how to plug the various gaps on my list. This resulted in some situations which in hindsight seem rather quaint, like the agonising wait in the observation room at the RSPB's (Royal Society for the Protection of Birds) Vane Farm Reserve for a little egret—at the time still a rare bird—to appear from behind some bushes. I also remember the utter despair of failing to see a red kite that had taken up temporary residence on the Scottish east coast not far from where I grew up. Suffice to say, in hindsight my anxiety about these two birds seems absurd given how regularly I have seen both species in the intervening years.

But this angst-ridden, voracious approach to birding showed no sign of abating as I got older. I soon became interested not only in maximising the length of my British list, but in the wider dynamics of the birding and twitching scenes. Even more embarrassing than my paroxysms of little egret-induced stress, was the strange mixture of jealousy and admiration that I exhibited towards high-profile members of the UK twitching scene. I hardly knew any of them in person, but that was no obstacle to me affording some of them an almost mythical status that I aspired to one day emulate. Fortunately, my more extreme twitching ambitions were thwarted by a combination of not being able to drive, and going to university, where my obsessive birding impulses were soon counterbalanced by the more standard temptations of undergraduate life. But they were not extinguished entirely and resurfaced somewhat in later years. However, resurgent aspirations towards UK twitching were dampened—literally and figuratively—by my experience of the famous Sussex great spotted cuckoo in 2005. Sure, I saw the bird; but as I stood in the rain in Brooklands Pleasure Park—every bit as unpleasurable as that sounds—I thought it time to re-evaluate my relationship with birding. This was given further impetus by my increasingly frequent visits to mainland Europe, which offered salutary reminders of an obvious but much neglected fact in UK twitching: birds we call rare are not, strictly speaking, rare. They are merely displaced individuals of globally common species. I was struck by the contrast between the amount of time and energy I had invested in twitching a serin at Rainham Marshes in 2009, and the ease with which one could find them in abundance on the near continent.

After a string of incidents comparable to that involving the serin, I gradually lost interest in what I considered to be the parochialism of UK twitching, and focused my attention on birding overseas. I embarked on a long-term project of trying to whittle down the number of regularly occurring European species I had not seen. And overall this was pretty successful, leading to some memorable holidays. But it was still rather stressful much of the time. When you travel far, often paying considerable sums of money for transport, accommodation and guides' local knowledge, the stakes can seem even higher. And although I saw lots of birds and had lots of fun in places like Brazil, Bulgaria and Belarus, something about the ethos of these holidays troubled me. Rather than a more rounded engagement with the cultural and historical interest that these countries had to offer, these were often smash-and-grab birding missions.

At precisely the time I was starting to grapple with these concerns, I came across discussions of low-carbon birding on twitter. The argument being made—that it is both feasible and desirable to cultivate a less carbon-intensive approach to birding and nature conservation—was hard to refute. And yet it made me uneasy: I was deeply invested in the ritual of my annual birding holiday, and I still had a number of destinations and species that I wanted to tick off. But my initial unease wore off thanks to two main factors. The first was a holiday to the Swiss Alps, my first (and so far only) experience of low-carbon overseas birding. Logistically, it was straightforward. I was easily able to travel from my home in Coventry to my final destination, the ski resort of Zermatt close to the Italian border, by train in a single day. Over the space of two and a half days I managed to find an excellent range of Alpine species, including snow finch, Alpine accentor, rock thrush, rock bunting, Alpine chough, nutcracker and citril finch (although I concede my wallcreeper efforts drew a blank). These were mostly seen on foot from my hotel, although I required the aid of the mountain railway to ascend to the altitude necessary to find snow finch. It was clear that high-quality overseas low-carbon birding may not be as troublesome as I had initially assumed.

The second key factor that drew me into more serious low-carbon birding was, as for many people, the Covid-19 pandemic and the assorted restrictions that came with it. The various lockdowns compelled me to embrace local patch-oriented low-carbon birding to a much greater extent than I might otherwise have done. Living in suburban Coventry, I do not—by any stretch of the imagination—live in a particularly bird-rich area. But the pandemic made me realise that I had previously underestimated the birding interest to be had on my doorstep. In 2020 and early 2021, I managed to find a yellow-browed warbler, a dipper, and two firecrests, all within a mile of my house. Granted, these are not national rarities, but they are all local scarcities and generated considerable local interest. And they would have gone unnoticed had I not taken up local patch birding to the extent that I did. But embracing low-carbon birding was more significant than simply becoming more aware of the avian delights on my doorstep. It also made me more appreciative of my local area. I only moved to Coventry in 2014, and my partner and I had our first child shortly thereafter, meaning that I was too engrossed, initially, in the early stages of parenthood to properly settle in the local area. Sustained local patching has afforded me a sense of connection to my local area that was previously lacking. The pandemic has obviously been terrible in many respects, but I suspect I am not alone in feeling that, on balance, its impact on my birding has been positive.

In embracing low-carbon birding, trying to conduct one's birding in a way that aligns with our knowledge of the severity of the climate crisis is clearly a major consideration. But it brings with it other beneficial consequences. For one, the turn to low-carbon birding has helped generate some much-needed introspection about the wider norms and values that shape the birding and nature conservation communities. And this has to be seen in the context of wider ongoing debates about equality, diversity and inclusivity in birding and nature conservation. Indeed, I have been gratified by the fact that—in the aftermath of #MeToo and Black Lives Matter—there are serious and ongoing conversations about how race and gender shape the norms and habits of the birding community, often in ways that make birding less appealing for people from marginalised groups. For me, low-carbon birding has to be part of that same conversation. It must be seen as part of a collective effort to imagine and implement a birding culture that is more diverse, inclusive and sustainable than before.

Furthermore, perhaps the most direct impact of low-carbon birding for me has been the evaporation of the stress and anxiety that has accompanied most of my birding career. Although I thought my shift from UK twitching to overseas birding was a major change, in hindsight it was in fact a continuation of the same anxiety-inducing logic that underpins much of the wider discourse about birding in the UK: that is, a deficit model of birding, premised upon obsessing about what we have not yet seen. This model of birding—a voracious, all consuming desire to see more and more species of birds—is still dominant despite the obvious harm it causes: it breeds competitiveness, stress, anxiety, and neurosis. It can harm an individual birder's mental health as well as being a recipe for domestic strife. And, obviously, it is not conducive to lowering one's carbon footprint. That is not to say that the enjoyment of rare birds per se is a problem: I like rare birds as much as anyone, and I often speculate as to which rare birds might plausibly be found on my local patch. But I do think we need to change the disproportionate focus of UK birding culture from the manic accumulation of sightings—and, increasingly, photos—of rare birds (or, rather, displaced individuals of globally common species), and shift the balance in UK birding towards more sustainable, inclusive, and environmentally friendly behaviours.

That said, I am not naïve about the uphill struggle faced by those committed to a lower-carbon birding culture. For example, a constant challenge is the frequent depiction of low-carbon birding by its critics as joyless and puritanical. For me, low-carbon birding is not about completely renouncing all my earlier desires and habits. I will certainly significantly reduce my overseas birding travel in future, but I have not committed to stopping entirely. And I fully concede that I feel unnerved by the fact that a genuine commitment to low-carbon birding may mean forgoing the chance of seeing certain species I have long wanted to see: my dreams of seeing ibisbill in Central Asia may remain unfulfilled! But I have already identified places in Finland, southern Spain, the Netherlands and Germany which have good birding possibilities and can be easily reached from the UK by train.

Above all, the discussion around low-carbon birding makes visible something which many have long known but never really confronted: that there is a disconnect between the culture of birding, and the need to develop habits and behaviours that limit, rather than hasten, biodiversity loss and climate catastrophe. We all know that if there are not major changes, we are heading for a real, extreme climate emergency. And yet we carry on as before because others do likewise. A necessary condition for the avoidance of further climate catastrophe is the promotion and normalisation of habits and lifestyles that do not rely on high-carbon emissions. Some may think this means abstinence and sacrifice, but my experience of low-carbon birding suggests otherwise. As my trip to Switzerland testifies, it is perfectly possible to have rich, varied and rewarding birding experiences even if you completely give up flying or travelling by car. My experience of local patching in an area not traditionally known for its prime birding also suggests, to me, that many of us underestimate the birding possibilities on our doorstep, in part because we tend to habitually converge on well-known birding hotspots. There is no denying that in embracing low-carbon birding I see fewer rare birds, and fewer birds overall, than before. But, ironically, the levels of enjoyment and pleasure I derive from my day-to-day birding have increased dramatically. Sadly, my grandmother-in-law died shortly after her one hundredth birthday earlier this year, but I take some comfort from the fact that, in her last year or two, her perception of my passion for birds was, finally, not so wide of the mark.

23

Redrawing my Birding Horizons

Sorrel Lyall

The prospect of exploring other countries and continents and immersing myself in a new world of wildlife motivates, fulfils and inspires me. But how can I justify carbon-costly air travel when I care about the environment and the wildlife that depends on it? How can I justify this when we are in the midst of a climate crisis?

For the past few years, this internal battle has been bubbling to the surface. I must admit that I have pushed it to the back of my mind for far too long and only recently have pledged to stop flying. Yet, as a birder and nature lover I long to see more of the world and its wildlife, so I have had to come up with some solutions and reframe how I think about engaging with nature.

For now, while I am tied to the UK with university studies, I will prioritise local nature and engage more deeply with the wildlife on my doorstep. I have never been a dedicated patch birder, only visiting the same sites every so often. However, when Covid-19 lockdown hit in March 2020, local birding took on a new meaning and became a great source of comfort. The day after lockdown was announced, with a daunting amount of free time on my hands, I took my daily exercise to the woodland down the road and almost instantly felt calm. It was the drumming of a great spotted woodpecker that reassured me; if the birds were getting on with life, I should too. And what a gem that woodland was, with dippers and grey wagtails along the burn, and bullfinches and treecreepers flitting above. Observing the woodland's changes throughout spring gave me focus and stability. Seeing the last redwing foraging in the undergrowth, hearing my first chiffchaff, then blackcap, then willow warbler, noticing the wood anemones appear and the marsh marigolds… watching the cycles of nature just getting on with it helped me to do the same.

Based in Edinburgh, I am very lucky to have Musselburgh just a cycle ride away. Watching the winter roll into spring here is an exciting time. As pink-footed geese fly north overhead, scanning the sea yields long-tailed ducks fading from their breeding plumage finery, while the orange tufts start to peek through on the Slavonian grebes. The white nape of the regular male surf scoter among the velvets catches your eye even at a distance, but when the velvet scoter raft comes in close to the sea wall the views are delightful. The wader scrapes are graced by bar-tailed godwits, knots and redshanks at high tide, and the surrounding scrub gains more life as the sedge warblers, 'groppers' (grasshopper warblers) and whitethroats return. Absorbing all the sights and sounds of spring at spots just a bike ride away from home has given me a far greater appreciation for local birding.

Getting involved with local monitoring work has also brought purpose and a sense of responsibility for the nature around me. I have always said birding is the only thing I will happily wake up for before dawn. Catching the sunrise on the way to ringing or undertaking a tetrad survey is hugely rewarding, as is contributing to important monitoring work. This is something I hope to take forward when I move away from Edinburgh—closely observing the bird populations around me and seeking fulfilment from surveys, ringing and regular patch birding.

I have also realised in the past year that the birding experience for me is as much about the people I am with as the birds we see. The awe of a close encounter or chancing upon some unusual bird behaviour feels all the more special with friends, family or a group sharing that infectious excitement. Visiting local sites with the university birding group in the past two years has been wonderful because I have found like-minded individuals after previously feeling weird and as though I stood out. It is birding with others that has made staying local that much more fulfilling.

So that is what I intend to do for now; but after graduating I plan on leaving the UK and living abroad. My goal is to travel slowly by boat, train, or bus, and stay in each place for a while and work along the way. These plans may not be easy to fulfil—with challenges related to work, money, time, and life commitments—but the challenge excites me, as does the thought of all the places and wildlife I will see on the way. And in every place that I live, I will engage with my local sites. I will get involved with surveys and bird them regularly with friends. I will join groups of like-minded people, all appreciating the birdlife around them.

We know that we need to start acting now to reduce the impacts of climate change. Although the onus is on big corporations and governments to make large-scale changes, as an individual I want to contribute to reducing demand for air travel, plastic packaging, and consumerism. Staying local and slow travel are part of a wider change towards a more sustainable lifestyle. I became vegetarian four years ago and would now call myself a flexible vegan (or 'cheegan', as I have been called). I think it is important to know such lifestyle decisions do not have to be all or nothing—just doing something helps. I buy packaging-free produce when it is within my budget, I buy clothes from charity shops and repair clothes with my limited sewing abilities. I have been practising these changes for a few years now, so it is time that I pledge to reduce my carbon-costly travel as well. I still drive a car, as it is pretty essential for ecology work, but I use my bike for local trips and love cycling down the scrub-lined bicycle routes in Edinburgh. But flying for a birding trip will no longer be an option I choose.

The climate contradiction in birding has always baffled me: wanting to conserve and protect natural places yet wanting to travel and fly around the world. The solutions are not straightforward and do require being open to reconsidering how we enjoy travel and how we want to engage with places, but I think every birder can find a solution that works for them.

24

Island Holidays by Train

Amy Robjohns

From time to time it can be good to have a change of scene and explore somewhere different from the local patch. This is even more the case when there is a chance to see new or different species. Scotland has become a favourite, particularly the Scottish islands, providing a level of peace and escape away from busy urban areas. Oban, Aberdeen, and Mallaig are easy to reach by train, and from there most of the islands are just a ferry-ride away.

One such visit to Scotland in 2019 was a solo trip in search of corncrakes, a species high on my wishlist. The Hebrides looked like the best place to try and after some research I settled on Iona, a small island off the south-west coast of Mull. Why Iona? It is a tiny island, about two miles by one mile, straightforward to access, with a wide range of accommodation available, and it is easy to cover the whole island in a day or less. Importantly, it also hosts breeding corncrakes.

While making the final arrangements, reports surfaced of a king eider at Nairn and I figured a 'slight' detour could work... On the evening of 29 April, I began the journey north—train to London and underground to London Euston before boarding the Caledonian Sleeper to Inverness. After arriving just before ten in the morning, a 15-minute train journey to Nairn and a short walk through the town was all that remained before I could embark on the tricky task of finding the eider. Nairn was lovely. There were plenty of birds to enjoy, and dolphins! Eventually I found the eider flock and, sifting through them, managed to pick out the king eider. There were also many long-tailed ducks, great northern, red-throated and black-throated divers, and Arctic and pomarine skuas.

The aforementioned 'slight' detour meant taking a long journey: from Nairn to Inverness, Inverness to Glasgow Queen Street and, finally, on to Oban—about eight hours by train—ready for an early morning ferry the following morning. But the scenery was lovely, and it was a good opportunity to begin reading Benedict MacDonald's *Rebirding*. I awoke to fog and low cloud but was excited about the next stage of the journey. Before taking the ferry, there was time to enjoy the black guillemots around Oban harbour; they are not a common bird down south.

The crossing to Mull was smooth, with a small number of seabirds to note, including great northern divers, though it was a shame not to have good views of the island due to the fog. West Coast Motors runs three bus routes on the island and by chance the first bus to Fionnphort was the same bus used for day trips, so the journey came complete with free commentary. Finally, after one last ferry across the Sound of Iona, I arrived on Iona.

I had read online that a good site for corncrakes on Iona was behind the fire station—about two minutes' walk from the slipway—in theory. Alas, no sign. A local on the ferry had warned me that the corncrakes had not arrived yet. The northerly winds were not ideal, and migration had slowed. It was 1 May, though, with plenty of hours for exploring… After dropping off my rucksack and enjoying a delicious hot chocolate in the Iona Heritage Centre café (highly recommended!), it was time to go searching.

The staff at the café suggested that I head east towards the hostel, as one of them had heard a corncrake the previous night. Twenty minutes later: *crex crex, crex crex, crex crex*— hooray! Two males were calling from a garden adjacent to the hostel, and one sounded frustratingly close but, typically, was hidden from view. I waited by the gate, hoping it might pop its head out. As I did so, the owner came over, explaining that he had seen corncrakes multiple times and inviting me into the garden. Moments later, the closer corncrake appeared, calling, and wandered around the lawn. This was fantastic to watch—I had not expected to enjoy such a good sighting. Later, that corncrake walked under the gate, and annoyed photographers by getting too close for them to focus their cameras.

Feeling satisfied, I went off to explore more of the island, and continued to do so the next day. I was fortunate to enjoy a spell of beautifully calm weather and fortune continued to smile on me as I stumbled across a third corncrake, sitting on a drystone wall. Corncrakes aside, it was also lovely to see flocks of twite (another rare species down south), numerous wheatears everywhere, willow warblers aplenty, various passage migrants and so on. Dun-I, the highest point on the island, gave me panoramic views of the whole island, Mull, Tiree and Coll.

Sherryvore Bed and Breakfast is located at the most westerly edge of the island and is a well-placed base for a birder, with views of the golf course and bay. This appeared to be the best bay for a variety of waders. It was also just a short walk from the only loch on the island. This bed and breakfast was my base again when I returned in September 2019 and it became my sea-watching hide, from where I enjoyed hundreds of Manx shearwaters, skuas and divers. Golf course highlights included golden plover, wigeon and, best of all, ruff—I am told the first for many years! Another highlight during the return trip was an otter in the Sound of Iona, swimming close to the ferry.

The slight downside of my return trip was stormy weather—strong winds (around 30–50 miles per hour) and rain every day. Iona has many sheltered spots, thanks to the rugged nature of the island, and good waterproofs are effective so it was still enjoyable despite the wind and rain. My journey home was delayed due to cancelled ferries resulting in missed train connections. Thankfully, though, we were picked up by the passing Tiree and Coll ferry and I was able to make last-minute arrangements to take the Caledonian Sleeper, but I arrived home about 12 hours later than planned. My advice: do not book advance train tickets if your journey also involves ferries, and be prepared for alterations. None of this has put me off future adventures, and my next trip to Scotland is in the pipelines.

Seabirds also rank high on my wishlist, and I have made many sea-watching attempts to see them from my local patch at Hill Head, gazing out into the Solent and towards the Isle of Wight. A good sea-watching day would be poor for most sites, so the Scilly Pelagics caught my eye. The Isles of Scilly are also straightforward to access by public transport, and easy to wander around on foot. In August 2021, I finally had the opportunity to visit after Covid-19 restrictions eased. I was excited to escape to another small island, but I also felt a little unsure travelling, given the circumstances.

The UK's only other sleeper train service, the Night Riviera, arrives at eight in the morning after leaving London Paddington just before midnight. Great Western Railway offers a travel package including either the Scillonian or Skybus, which makes travelling to the island via public transport good value. I had decided to test out the tiny Skybus which takes only 15 minutes each way. Remote islands depend on small aircraft, especially when the sea is too rough for ferries. I had never flown before. I look forward to the day these aircrafts run on electricity, which according to experts may happen within a decade (unfortunately there is no prospect of electric long-haul flights at least for the next 30 years).

The majority of this trip was spent on a small boat in the Atlantic Ocean, up to 16 kilometres off the islands. The Birder Special weekend consists of two evening trips—one on the Friday and the other on the Monday, with two full days over the weekend. It did not disappoint. Not long after setting sail on the Friday, we picked up a grey phalarope, enjoying close views for a short while. Throughout the four days, Wilson's and European storm petrels provided brilliant views as they fed in the slick nearby, allowing for good comparisons. Shearwaters were trickier but we did have one great shearwater follow the boat, a brief sight of a Balearic shearwater—sadly, a rapidly declining species—multiple sooty shearwaters and, of course, good numbers of Manx shearwaters, which breed locally. The swell added to the challenge, especially on the Sunday when strong southerlies and a swell that reached three metres at times made standing difficult and much spray inevitable. On the Sunday we came across a mixed raft of Manx and great shearwaters. Alas, Cory's shearwater, one of my targets, failed to appear. However, the boat trips were still well worthwhile for the experience and to enjoy other seabirds at close range— beating distant scope views by miles. Common dolphins also made an appearance most days, coming alongside the boat; that is another species I rarely see in Hampshire.

Exploring via public transport can be a challenge. Some reserves are in remote locations many kilometres away from public transport, and not all of these reserves link up well with footpaths, cycle paths or pavements. Walking on roads is not particularly fun or safe, though cycling can help to link train stations to more distant reserves. Thankfully, there are still a good number of reserves and wild or pleasant coastal areas within easier reach for those without cars. One can even visit the Cairn Gorm mountain via a bus from Aviemore. Train delays and cancellations are also less than ideal, but so too are traffic jams, frequent road traffic accidents, closed roads and the long detours that accompany them. Public transport also has the added advantage of allowing you to be able to relax more than when travelling by car, as you enjoy the scenery from the window, read a book or even sleep while being transported to the target destination.

25

Lammergeyers from Leeds

Jonnie Fisk

In January 2019 I flew to Hungary with one of my best mates. We hired a little Suzuki Swift and toured around the snowy steppes and mountains of Eastern Europe under the pretence of paying homage to the wintering long-eared owl 'experience' in Kikinda, Serbia. I returned to my home in coastal East Yorkshire having had the usual formative escapade one does when travelling with good company to new places and seeing new birds: my first saker falcons and *caudatus* long-tailed tits, picking out red-breasted geese among thousands of white-fronts, experiencing a brambling roost that supposedly numbered into the millions and so forth... It was fun to burn rubber across four countries, spotting great grey shrikes, golden jackals and crested larks in the snow as we went. But was it right that the budget airline fare from Doncaster to Debrecen was less than half the price of my usual train journey from Hull to my hometown of Harrogate? The bargain flight, cheap accommodation and massive meals for under £20 left little reason to dwell on the environmental impact of what was essentially a quick winter jolly to see some owls.

With a year to sit on it, and crushing environmental anxiety from the daily dose of climate crisis news dished out every time I checked online or listened to the BBC World Service, I reached a small personal epiphany. In X number of years' time, in a warmer world with society reshuffled and my friends' children grown, I want to be able to look at myself and know that I made changes to my lifestyle in the face of undeniable and catastrophic climate change. We are constantly told of alterations—of varying importance—we can all make to play our part. I have heard that there are two biggies which considerably shrink your personal carbon footprint. One is switching to a plant-based diet—achievable (and I am getting there now) but undeniably requiring concerted effort on a daily basis, and varying in difficulty depending on where you live or who you live with. The other is cutting out air travel. This is as easy as literally not getting on a plane. I try not to preach or judge others outwardly on their travel, although this is getting increasingly difficult, but from that moment I knew that going forward, if I scratched that birding itch abroad, it would be without the assistance of an aeroplane.

I am extremely lucky that I live an inherently, even intensely, birdy life working on the Spurn peninsula, meaning there is often very little motivation to take myself elsewhere. But, barely having digested my Boxing Day leftovers, in December 2019 I found myself sitting on an overnight coach travelling from Amsterdam to Berlin, my binoculars tucked inside my hoodie and a familiar 'herbal' smell emanating from the other young passengers. By the morning we were in Germany, and less than 24 hours after I had set off on the train from Leeds, I was watching Berlin's iconic urban goshawks.

This trip I took across the near continent to Berlin, then to Italy as far south as Naples, back north to the Swiss Alps and home to Spurn by way of Amsterdam again, was not my first foray on Europe's rails. Fresh out of school, about seven years before, two buddies and I travelled from the Netherlands to Turkey for three weeks using an Interrail ticket—a rite of passage for scores of European youths on their holidays. The ticket cuts down costs massively and, in 2019, at 24 years old, I was still eligible for one. For some continental rail trips, buying individual train tickets works out cheaper, but if you want some wriggle room on dates and locations, greater freedom of movement, or are planning on packing in lots of journeys over the selected time period, as I was, an Interrail ticket is a no-brainer.

I have visited Berlin before in the summer, also via train, and knew how easy it was to spend all day lost in the Großer Tiergarten with icterine warblers, striped field mice and red-breasted flycatchers. But winter brought a whole different experience and during my three days there I would walk the 30 minutes from my hostel and spend all daylight hours in the park, not really sure what to expect apart from the goshawks I had come to study. Flocks of hawfinches, bramblings and nominate bullfinches fed in the sycamores, and one afternoon a skein of white-fronted geese flew over low. At least three middle spotted woodpeckers were faithful to particular areas of the park, looking pink-flushed and full-bodied in the frosty foliage. I watched them for hours.

I have seen goshawks before, sure, as barrel-chested silhouettes displaying over Breckland or upland clearings. But there is something about the promise of seeing one of these battleaxes of the Palaearctic's primeval forests in a city park, with joggers and prams and defecating dogs passing under them, which drew me back to Berlin. And it is just as described. The goshawks preen, feak specks of flesh off their chops, and just generally stare wild-eyed into a private world of murder and meat, while city life carries on beneath them. The huge flocks of woodpigeons in the park feed restlessly, constantly looking over their grey shoulders in a way you do not see in Britain. A goshawk culture breeds an untamed air in these urban parks: a carnivorous iron-grey agrestal weed among labelled, imported trees and cyclamen borders. A wildness missing from our nation's cities—but only a bus ride away.

I should say now that long-distance train travel (especially in Europe) is, generally, marvellous. Rather than getting from A to B in a soulless, airless aeroplane cabin, viewing all the places in between your destinations as circuit boards of cities and capillaries of rivers from 42,000 feet, you watch each vista slip slowly into another, note the gradual changes in landscape use, in architecture, in the flora and—as with this journey—the crows turning from carrion to hooded. Not only is it visually pleasing but it is very settling for the mind. Legendary travel writer Paul Theroux detailed in *The Great Railway Bazaar* how 'Train travel animated my imagination and usually gave me the solitude to order and write my thoughts. I travelled easily in two directions along the level rails while Asia flashed changes at the window, and at the interior rim of a private world of memory and language.' There are very few activities I find as cathartic as notebook admin: fleshing out sketches, transcribing conversations and backlogging birds on A6 pages. Sitting there on the Munich–Bologna service, munching happily on a *Bündner Nusstorte*, was catharsis squared.

Of course, cities are easier to reach on trains and buses than prime European birding sites. But one can compromise. After Berlin, I celebrated the New Year Italian-style with an old work colleague in Lombardy before we spent the first day of 2020 travelling by rail to Rome. I expanded my culinary horizons and attempted some 'culture' there with my native guide but will admit I was mainly focused on the blue rock thrushes, black redstarts and serins around the Roman ruins, lesser spotted woodpeckers and nesting monk parakeets in the city parks and, come sunset, a distant—but still breathtaking—dance of a million or so starlings over the city skyline from Castel Sant'Angelo. One of the joys of our modern birding intranet is the ease with which the prospective traveller can gen up on unfamiliar sites, glean morsels of local knowledge and then add to the collective experience for the next birder in transit.

It was then onward to Switzerland, my belly full of pasta and fingers itching to exercise on the focusing wheel of my binoculars in 'the wild'. Even Alpine species are not beyond the range of rail and bus birding—if you save a few spare francs for a ski-lift ticket. A couple of hours' travel out of Zürich and I was crashing through the snow of the Gemmi Pass, where Alpine accentors descend upon any discarded crumbs. The Gemmi Pass is accessible from the ibex-skull adorned streets of Leukerbad, and the cable car up provides an excellent vantage point from which to see chamois at play. The remnant populations of these rock-jumping ungulates provide carcasses for bearded vultures. Watching the shaggy angles of these bone-guzzling gargoyles turn in the air below is a pure *event*, and a lifetime goal was achieved when a lengthy shadow slid over my back as I threw more nuts at another scavenging Alpine accentor—an experience straight out of a *Collins Bird Guide* vignette.

Several hours to the east, the Bernina railway sends charming red trains crawling over montane bridges, across frozen waterfalls and through the chocolate-box villages of Pontresina, where I disembarked and wandered around the pine forests, occasionally leaping out of the path of wild cross-country skiers. Willow and crested tits were easily coaxed down from the branches and into my seed-stocked palm. 'Black' red squirrels and spotted nutcrackers showered snow down as they moved through the beard lichens. It was fairy tale birding for a lad from East Yorkshire's muddy agricultural flatlands.

My home set-up, at sea level between the brown North Sea and browner Humber estuary, has conditioned me to find these mountainous landscapes oppressively beautiful. Mythical birds deserve suitable environs, and another Swiss public transport birding day ended on an olde worlde bridge, separating the birder from an icy death down a glacial ravine by just a metre of so of rock. A wallcreeper watchpoint, accessible by a small, montane bus service. No finer place to spend a winter afternoon as the light wanes. I waited an hour or two, enjoying the gradual desaturation of the blue-and-orange resident rock buntings as the sun dipped behind the peaks until… there!… that miniature, bounding flight, those floppy, claret wings. A wallcreeper. Some birds just cannot be overhyped. The geological processes of the setting were rendered insignificant in the presence of this flamenco Sherpa nugget glued to the vertical like some holy fridge magnet.

I have experienced the demise of rural bus services in the UK, as their demand wanes and budgets are cut, a situation starkly juxtaposed against the Swiss public transport system, where bafflingly discreet bus stops, marked on painted boulders on nondescript mountain roads, allow you to catch a ride deep into the valleys to kick start a day's hiking or hail a people-carrier-sized village bus back home after watching *Tichodrome* magic until dusk bruises the white mountain ranges purple and blue. There will be numerous easier sites where wallcreepers winter in the Pyrenees, or as unpredictable waifs wintering on European architecture closer to the Channel—all accessible without a plane, or even a car in some circumstances. Before we tackle the issue of whether we should travel to see these birds, you can at least, for now, address how you travel to see them.

Yes, of course, birding abroad without air travel or a car requires more effort. It requires planning. It requires time. It sometimes requires long walks along hairpin Alpine highways, with ravens bounding along behind you. But perhaps a shift towards a slower pace of movement, via overland means, spending longer at each place, is a step in the right direction for those that still want to travel while conscious of their carbon budget. Productive, fulfilling, travel incorporating

European birding is out there—and an Interrail pass makes this easy and affordable.

You would be naïve not to realise that travel for birding itself is a privilege, and the climate emergency should provoke all of us to redefine what is regarded as a legitimate and ethical way of enjoying birds. Should I travel at all? My day-to-day life certainly is not ornithologically impoverished enough that I need to seek out birding therapy abroad, or even elsewhere within my own country. But it is comforting to know that if I want to experience some continental birding, one of these tickets, some prior planning and an open mind is all I really need.

26

Bringing Birding Home

Nick Acheson

Life is all about irony, and nowhere more so than in our relationship with nature. In the UK we cast ourselves as nature lovers when—demonstrably, if we would only stop to look— we have harried nature so successfully from the landscape that it cowers only in forgotten corners. There is a double irony to being a naturalist and, more particularly, a birder: sometimes the very act of watching the birds we love is destructive.

My own story with birds is one of extraordinary good fortune. I grew up in North Norfolk and went to school in Holt, where I was taken under the wing of a wonderful biology teacher (still among my best friends) with whom I would make the short journey to Cley each week. Ours was a school birdwatching club for which semipalmated and broad-billed sandpipers, tawny pipit, Leach's storm-petrel, golden oriole, red-footed falcon and Montagu's harrier were perfectly normal birds to see on a Saturday morning. In the third year of my undergraduate degree I moved to southern France, between the Camargue and Crau, where I fell deeply for pin-tailed sandgrouse, little bustard and southern grey shrike. At the end of my MSc I moved to Bolivia where for a decade I worked on a range of projects related to wildlife and conservation.

It was only by accident that I began to lead wildlife holidays. I worked in remote northeast Bolivia, in a little-explored national park, where I saw my first jaguar, giant anteater, giant armadillo, maned wolf, pink river dolphin, and countless birds of Amazonia and the Brazilian Shield. Because I spoke Spanish and English, and happened to know the wildlife, I was asked by park authorities to lead a visiting group of donors. I did, and then I was asked to lead more. Next I was asked by overseas companies to lead their birding groups round Bolivia. When I left Bolivia I was asked to lead elsewhere, so I suggested India, which I knew well. This went wonderfully and soon I was working right around the world, on seven continents, sharing wildlife with people.

Though leading bird tours wholly cured me of listing, I saw a huge chunk of the world's landscapes, seascapes, mammals and birds, and I felt richly privileged. I also amassed a vast carbon footprint.

Wildlife tourism is discussed elsewhere in this book. My aim is not to defend the fact that for years I roamed all over the globe and—in return for looking after sometimes very demanding clients—I had an absurdly privileged relationship with the world's wildlife. I often fretted over my heinous carbon debt, but persuaded myself I was doing the right thing, with all the usual arguments rehearsed in favour of wildlife tourism. I wanted my tours to achieve more for wildlife and shifted to a company I saw as greener and more ethical, asking that my tours should directly raise money for conservation. I excused myself, too, with the—true, but self-deluding—fact that, as a ruthlessly minimalist vegan, my environmental impacts were limited when compared to those of many westerners.

I need to say here that I have decent, ethical friends who still believe that wildlife tourism is a tool in the preservation of nature, a significant contributor to the economic defence of wild spaces and wild species, including some friends who have devoted their entire careers to conservation.

I can see their point. However, the weight of my actions came to crush me more heavily with every tour. I would sit on yet another long-haul flight, watching hundreds of people tearing into plastic packaging, using disposable items for mere seconds, while collectively we pumped literal tons of carbon into the atmosphere. We would then stay in temples of consumerism, often wildly at odds with the poverty of a nation's people. We would casually board internal flights to reach remote—allegedly pristine—wilderness, in order to see, to experience, to persuade ourselves we were alive.

I broke at last when I began to read more seriously about the climate crisis. Naïve as it sounds, I have always tried to align my actions with my ethics. I declared as a child I would be vegetarian. I later became vegan. I have volunteered and worked for wildlife charities since my teens. And here I was presented with the blunt, brutal fact that we are hurtling at ever-greater speed towards a climate crisis, to which I was contributing, which would—inarguably—tip our world into famine, extreme weather, unrest, desertification, mass migration, habitat destruction and species loss on a scale we have barely begun to conceive. The facts are clear, the scientific models speak in unison: unless we drastically and immediately cut our greenhouse gas emissions, biodiversity as we know it is done for, and human life will be immeasurably crueller, especially for those who are already vulnerable.

It is always the redeemed convert who is the most irritatingly zealous. In the space of a year I gave up leading overseas tours. I could no longer bring myself even to drive the short distance to the coast to watch wildlife. Then—among the many ironies of 2020—the Covid-19 lockdown came. It kept me forcibly in one place, on foot, for months, during which I witnessed a miraculous spring. Though ten miles inland, for two weeks I saw ring ouzels every day. I sat late at night listening to teal, wigeon and common scoter migrating over my garden. I walked in the dark, hearing tawny owls, barn owls, little owls and woodcocks. One of Roy Dennis's white-tailed eagles flew over my house with a buzzard up its tail. One of Pensthorpe's corncrakes settled in a meadow nearby and could be heard from my bed. All ten of my locally breeding warbler species brought their voices back to my little Norfolk valley. When strict lockdown ended, I had my mum's 40-year-old bike repaired and began to roam a little farther.

Despite close family shielding from Covid, despite being entirely alone for months, despite worries over work and income, for the first time in many years I was truly home. Home among the birds and other wildlife with which I grew up.

Strangely though, of all my comings out—gay, vegan, treehugger, rescuer of hopeless animals—the one about which I most often feel judged—as woke, virtue-signalling, holier than thou—is giving up flying, giving up driving to see birds. Because it poses, I suppose, the biggest cognitive threat to a deeply entrenched way we have of loving wildlife: of acquiring ever-wilder fixes of birds and wildlife.

Yet if we continue to love wildlife in this way we will facilitate the death of biodiversity. If ever there was a time of reckoning—a time to accept that the act of watching wildlife does not itself equate to doing good for wildlife—it is now. It is beyond time that we lovers of the wild became witnesses of the climate crisis and acted in accordance. And that we brought our birding home.

27

Little Steps, Big Difference

Steve Dudley

Our planet's biodiversity is ultimately tied to the planet's health, and right now it is effectively in critical care state. Biodiversity and climate are inextricably linked and if governments continually fail to tackle both the biodiversity and climate crises, then our planet will soon be in such a steep decline that things will be irreversible. Some say we have already passed this point.

Without ambitious climate policies, and without legislation and infrastructures to enable all of us to live more environmentally friendly lifestyles, the onus remains firmly on the individual to reduce their own impacts on the planet while demanding system change. How each of us chooses to do this is dependent on many factors and not everyone has the same opportunities to reduce energy consumption, but even within current constraints our personal actions matter.

I used to believe that an individual's actions bore little impact on the fight against climate change. When weighed against industry or national impacts, sure, an individual's impact is miniscule. But the more I began to understand my own lifestyle, my own energy consumption, my own climate impacts, the more I realised that collectively, those who care enough to make changes, can make a difference. I guess, like most people, I was just too slow to wake up to the real impacts of climate change, what was driving it and how I was contributing personally.

I have been a birder for 36 years, and for over 20 of those years I twitched the length and breadth of Britain and enjoyed foreign birding trips around our wonderful planet. My increasing carbon footprint became a concern for me around 20 years ago, reaching a point 15 years ago where I felt, as a conservationist and environmentalist, and in the absence of effective legislation and our government's continued failure to meet agreed environmental targets, I had to do something myself. I could no longer consciously call myself either a conservationist or environmentalist while living the lifestyle I did.

Reducing the size of my birding carbon footprint had to be part of reducing my overall carbon use. There was little point reducing only a segment of my impact, but my birding was so fundamental to my lifestyle that it was where I had to start. So, I stopped national birding and twitching in favour of a local area—not a local patch, but the wider area of the Peterborough Bird Club which covers around 500 square kilometres. I also reduced my overall overseas air travel: instead of going on two overseas trips in a year, I went on one longer (in time) trip, reducing my air travel by half.

As a result my local birding intensified and, as my miles began to rack back up again, I had a rethink and further reduced my local birding area. This I have done over the last 10 years—my home area comprises my immediate fen (which I can walk out on to from my house), adjacent fens and two main sites each under ten miles from the house. This really brought my birding mileage down to what I was comfortable with, with an overall reduction from more than 60,000 miles per annum to less than 5,000 miles.

I am lucky to live in a productive birding area so the localisation of my birding over the last 15 years has not meant unexciting birding—far from it! The Fens are teeming with birds year round. My immediate arable fen still holds good numbers of declining species such as corn bunting, yellowhammer and grey partridge. Hobby and long-eared owl breed locally on the wider fen and our garden list stands at 150 species and is the best site in the county for tree sparrow. Wetlands within ten miles of the house hold breeding crane, bittern, marsh harrier, black-tailed godwit and garganey. In winter they are wildfowl and wader havens; during spring and autumn plenty of interesting migrants often turn up.

My international travel remained a concern. Not only was I still flying but I had been working within ecotourism as a bird guide for over 25 years. Reducing my bird guiding to a single location, Lesvos, was the first step. This meant giving up being paid to visit exciting countries and the income that went with it. I was embedded within Lesvos ecotourism, having written the birding guide for the island, but I could still change my own operations in order to reduce my carbon footprint while there. I stayed for longer periods and stopped taking groups out from the UK but switched to offering day trips to visitors already travelling to the island. However, ultimately, the Greek economic collapse coupled with the Syrian refugee crisis effectively cut around 98% of tourism on the island. I had little concern for myself at that point, my concern being for all my Lesvos friends whose income was at best reduced to a fraction of what it had been, and at worst the loss of everything they had worked for.

Back home, my wife, Liz, and I set about converting two small fen cottages into a single house in as environmentally sensitive a way as possible using (more expensive!) green alternatives for the build (for instance, wooden-framed windows, natural insulation materials, water harvesting).

Liz and I are committed to further reducing our personal impacts on our planet. As much as I love travelling the world and seeing new places, cultures and wildlife, I have decided to stop all international air travel. This has been a massively hard decision to make as I have made it while still being able to afford to travel anywhere in the world. So, I will never see a tiger, a blue whale, an albatross, and much, much more that I have long dreamed of seeing in the flesh.

While I accept that by no longer travelling to areas that depend on ecotourism I am depriving lodges and other local services of valuable income, global conservation should not be based on the affordability of a destination for those with expendable income. Such a system is ultimately flawed (as exposed during the Covid-19 pandemic), and while some nations have increased their conservation efforts and spending as a result of increasing ecotourism, just as many countries have done nothing. This capitalistic system concentrates ecotourism in areas with the most sought-after species and it ignores huge swathes of the planet because they do not offer such 'must see' species or they do not have the infrastructure to support even the most rudimentary of tourism demands of the twenty-first-century traveller—the majority of globe-trotting birders today expect to travel and stay in comfort, enjoy good food and be led by the best guides. After all, they are paying top dollar for their experience.

Liz and I have recently enacted our plan to retire to an island archipelago that already generates all its electricity needs using renewable energy sources—and one which is rich in birds and other wildlife year round. Being an inland-based birder all my birding life, the lure of coastal birding in retirement remained too great, and once Liz and I started searching for our birding retirement area, the appeal of island life and remote birding grew. Throw in the increased chance of finding your own rare birds, then moving to one of our migration and vagrant-rich island groups became a must.

Our move to Westray, one of Orkney's northern isles, contains us within just 18 square miles. It is an ideal place for us to switch to an electric vehicle (run on locally generated renewable energy) and we can do more bird and wildlife watching on foot and by bicycle—we have miles of footpaths and the few roads we have are reasonably traffic free and as safe as you could wish for cycling. With good ferry connections to mainland Scotland, we will replace overseas travel with exploring the wilder areas of Scotland (areas that we have yet to reach and favourite areas we will return to), taking advantage of the increasing network of electric vehicle charger points (many of them free in Scotland), as well as exploring our own island archipelago further. This move has also seen me give up another of my real, but carbon-hungry, passions: my Manchester United season ticket—and the 7,500 car miles that come with it.

I have also been able to make a small impact via my employment. For 25 years I have run the British Ornithologists' Union (BOU). As well as working from home for all of that time (no commute with associated carbon emissions), part of my job has been arranging face-to-face conferences and meetings (remember them?). When the BOU joined social media in 2011, I soon became aware of the ability of social platforms to bring together people from all around the world. Having helped to establish Twitter as the social media platform of choice within ornithology, I was keen to use the platform for global conferencing. The World Seabird Union was the first to take to Twitter to run a series of annual conferences from 2015, and the BOU followed in 2017. From 2019 I arranged for the BOU to support the World Seabird Twitter Conference series and in 2020, with the International Wader Study Group (IWSG), we ran the first International Shorebird Twitter Conference (#ISTC20)—something I am sure the IWSG will now make an annual event. From 2021 it has been BOU policy to run all their conferences as dual-format events, with a Twitter conference running parallel to each in-person or virtual conference they host.

Twitter conferences allow both presenters and attendees from all around the world to present their research and/or attend a truly global event; few of them would be able to attend an in-person event. These events are as low carbon as you can get. No travel. No catering. No running of energy-hungry venues. And they come with other important community and societal benefits, including increased inclusivity and diversity of attendees. They enable those with little or no funding to attend and/or present their research—something they could not dream of doing at a far flung international in-person meeting.

I continue to make efforts to reduce the carbon demands of my personal and professional lives. At the same time, and for as long as I can afford to do so, I will continue to support non-governmental organisations and others committed to researching and conserving our world's birds. So, it is not a question of turning my back on the need to conserve the planet's biodiversity, but rather controlling what I can personally and contributing to the global conservation efforts of the institutes best placed to fight on the biodiversity front.

Do not think you have to escape to a remote island to do any of this. Liz and I achieved most of our carbon reduction while living in fenland Cambridgeshire. Had we stayed we would have still have made our switch to an electric vehicle and dropped the more carbon-hungry elements of our lives, but retained lower-carbon activities (for example, flying and my Manchester United season ticket would have gone, but my Peterborough United season ticket would have remained as the ground is only seven miles away). Those living in towns and cities may have better opportunities than those living in rural areas to reduce carbon consumption. Not owning a car and using local public transport is a great carbon-reducer, but it should not stop you hiring an electric vehicle for day trips that are out of reach on public transport (even when combined with a bicycle), weekends away and holidays.

Liz and I surprised ourselves at how easy it has been to make the changes we have made to not just reduce the carbon demands of our bird and wildlife watching, but across most areas of our lives. We feel richer for doing so and appreciate our local environment and its wildlife much more because of our renewed focus on it.

28

Climate and the Cuckoo Calendar

Lowell Mills-Frater

The call of the cuckoo has long been one of the most familiar and widespread sounds in nature to be welcomed back each spring. It is perhaps because of its familiarity—evident from the many references to cuckoos in culture and mythology—that the loss of cuckoos from entire landscapes in England, Wales, and many areas of Western Europe, has been so deeply felt. For some, the loss of their local cuckoos has been met with what can only be considered grief, and their stories are related by Michael McCarthy in his book *Say Goodbye to the Cuckoo*. The cuckoo's decline is played out in the hard data too, recorded by the British Trust for Ornithology (BTO) and similar organisations internationally. Scientists remain uncertain of which factors play the greatest role in driving the trend, but land use change, food supply, climate change and the difficulties they cause for migration, are all candidates.

Climate and carbon—in the form of plant cellulose, the indomitable root of insect food-chains—are tremendous forces in the annual cycle of cuckoos. Cuckoos depend on the predictable arrival of spring in the Global North, and of wet seasons south of the Sahara, to bring consecutive bursts of plant and insect life, replenishing the large caterpillars on which the adults feed and igniting the northern songbird nesting season that raises their often-vilified offspring. A 2017 study by the University of Copenhagen's Professor Kasper Thorup and collaborators, showed that individual satellite-tracked cuckoos' movements and stops in successive areas on migration through Europe and Africa, closely tracked each region's peak in vegetation 'greenness', as measured from NDVI (Normalised Difference Vegetation Index) satellite imaging data.[1] In sub-Saharan Africa, the rains and the surge in green growth are significantly influenced by the Inter-Tropical Convergence Zone, a ribbon of interacting trade winds from the Northern and Southern Hemispheres that sways in its position throughout the year.[2] In Europe, the warmer days in April see the first grasshoppers hatch and the impressive caterpillars of moths like the oak eggar and the drinker emerge from overwintering. Meanwhile, dunnocks begin building their nests; meadow pipits—short-distance migrants—arrive and rapidly pair up on the moors and longer-travelled warblers do the same in the wetlands. These resources and ecosystem processes, on which cuckoos depend, are attuned in their timing to the day length, the temperature, the plants and the insects—steady ripples across a planetary pool.

What is the outcome for cuckoos when these waters are disturbed? While research on the species' regional declines, including my own studies during my PhD, has highlighted, perhaps, key influences of land use and populations of their insect food,[3] there are indications that climatic mechanisms are at play, both separately and in interaction with this food supply. In a 2018 study, Dr Alison Beresford of the Royal Society for the Protection of Birds (RSPB) and collaborators further examined cuckoos' and other migrant birds' relationship with seasonal greening in Africa. They indicated that, should greening events occur later than normal (as is projected in future Northern Hemisphere autumns by the study led by Thorup) or shift further south, then this could result in birds entering spring in poorer body condition or delays to birds arriving at their breeding grounds due to longer stopovers en route.[4]

In the Global North, a number of typical spring events monitored by phenologists (who study lifecycles and seasonal change) have been shown to have advanced in timing in response to warming temperatures by an estimated 2.5 days for every decade since 1970,[5] a total of 12.5 days earlier to 2020. This equates to roughly the full period of egg incubation for many insect-eating birds, who time their breeding attempts to meet the peak in insect emergence, especially the hatching of caterpillars such as those of the winter moth in oak woods. For migrant birds, this means pressure to return from migration and lay earlier (something resident birds might more easily manage) to still catch this wave, and among these migrants are key host species of the cuckoo, such as the *Acrocephalus* warblers (reed, great reed, sedge) and the redstart. The cuckoo, therefore, is also under pressure to return early, to ensure the first clutches of these birds' eggs are being laid when their own eggs are ready to be stowed alongside them. As cuckoos are projected in the future to be delayed or lacking fuel as they leave their annual stopovers in Africa, they may be limited in their capacity to arrive earlier. Indeed, existing data analysed by the late Professor Nicola Saino and colleagues showed that while cuckoos have barely matched the advancement of arrival of their long-distance hosts in Europe (5.3 days to the hosts' 6 days) they have fallen behind their short-distance migrant hosts' earlier arrival (14.6 days).[6] Capacity to simply target alternative host species that also arrive later is also limited, as individual female cuckoos lay eggs with a shell pattern that mimics the eggs of their discerning host species (grey speckled eggs for pied wagtails, patterned brown eggs laid alongside those of meadow pipit and reed warblers, immaculate blue eggs to fool Scandinavia's redstarts). Laying these eggs into nests where they do not blend in has been shown to often be met with swift ejection by the host. Further, female cuckoos appear to be specialised in targeting one host species over other, similar, species (for example, sedge warbler but not reed warbler). A picture is emerging of cuckoos whose targets are short-distance migrant or resident species increasingly 'missing the boat' on that first wave of nests with eggs into which they can sneak their own.

Climate change is also bringing an increased occurrence of extreme weather events, and this has already included some which have spelled disaster for cuckoos. Local spring and summer flooding has on occasion laid waste to reed warbler nest sites—the species on which perhaps the majority of UK cuckoos now rely to raise their young, in precarious reedbed nests. In autumn 2012, three male cuckoos from the BTO's cohort of satellite-tracked cuckoos travelled through Iberia on their way south and found the landscape in a state of unseasonal drought. It proved a more deadly obstacle for that year's class of tracked birds than crossing the Sahara, with all three birds' transmitter signals lost before they even began the desert stage. As rare or atypical occurrences, such events are part and parcel of undertaking long-distance migration. It is when the frequency of these events shifts from once a decade to once every few years, as a result of anthropogenic rapid climate change, that we must stop and think about whether this is in fact a conservation issue which we must set out to resolve. More frequent extreme weather events evidently stand to bring cuckoos under attack from *multiple* angles, under which their lifecycles would ensure they had little to no room for error.

Clearly, then, conservationists must confront climate change further than by simply researching it. In order to keep the global temperature rise below 1.5°C, a profound transformation will be needed, in both the economy and society around it. Birders, as dependent as anyone on the continued health of ecosystems for their wellbeing and sanity, can and indeed must be part of this change.

My belief is that low-carbon birding entails no sacrifice. It is perhaps the most fulfilling, accessible, beneficial (to health and wellbeing) and exciting form that birding can take. That it could go a significant way towards preserving local environments and global climate, is simply a bonus.

I have tried and succeeded in birding locally from my front door everywhere that I have lived: in city and country, at the coast and inland. Each has had its own unique character and advantages. In fact I believe it is entering this micro scale of variation in what habitats and birds are present, what is common and what is rare, that is the core beauty of birding locally.

Setting off from your door on foot or by bike, the birding begins immediately. Out in the air and with no engine noise, you can hear bird calls and songs over virtually all of the round trip. On foot, especially, your attention can be tuned in to anything you choose. The birds, the landscape, the wandering of your thoughts, a catch-up with a friend. A bike may be required if the patch is further afield. Even so, if you get a lead on a bird en route, you can halt, take detours, and backtrack with much greater ease than in a car. A folding bike in particular allows you to build the bus, train, car-club or car-share into your route, opening up county-level birding, and beyond, in a low-carbon vein.

Such flexibility means that what may have been part of your route to a birding patch becomes a patch in and of itself. Birding locally, or along routes that some may call 'slow ways', allows you a greater opportunity to be your patch's most frequent visitor. It is so often these regulars who are most likely to notice the successive tiny cues, the fleeting views of flits and flight patterns, that can lead to the discovery of a new species for the site; a new breeding pair, nest or roost; even a whole new patch of habitat (a flash on a flooded field, an overgrown railway embankment). Intensively working a local area has allowed me the time and opportunity to study the ornithology from so many angles: reading and reporting rings fitted to birds such as black-headed gulls; identifying calls of water birds, such as whimbrel and golden plover, passing over at night as I walk my dog; and monitoring the progress of nest attempts by a range of species as part of the BTO's Nest Record Scheme. This is part of a bigger picture wherein one becomes finely attuned to all aspects of local biodiversity, and therefore aware of specific threats to local birdlife and better equipped to join campaigns. Only those who are aware can raise awareness.

Along with these benefits, low-carbon birding presents the opportunity to join a community of others patch-birding in the area, from crossing paths while out in the field or through channels like social media. During the events of the Covid-19 pandemic and lockdown measures, my running correspondence with another birder in the area on all things ornithology became a lifeline which no doubt aided my mental health. I have become heavily invested in their green patch year list, and I am delighted when I turn up something new. Since 2020 I have logged sightings using the Cornell Lab of Ornithology's global recording app, eBird. This has not only helped me place my local birding within a world birding context, its capacity to generate bird data and statistics within a set radius of home has provided great motivation to me to head out and find that next new species locally—a fierce competition with myself and other local birders, month on month. In addition, it has highlighted the local patch-birding habits of many others in my county. But I find the biggest rushes I experience are often from the less rare but equally interesting 'patch mega' birds. Being close to central Newcastle upon Tyne, my day has been lit up variously by yellowhammer, little egret, spotted flycatcher, garden warbler and Mediterranean gull (this last bird sporting a colour-ring fitted in Poland). Most poignantly of all, in the short period of writing this chapter I took an early August morning trip on my bike to Newcastle's vast Town Moor, a green expanse larger than New York City's Central Park. It was here that, dragging my bike over the tussocky ground to the only stand of trees, I glanced up at a flyover bird at some height. Looking at it through binoculars, it appeared to be equal parts pigeon and hawk. It was a juvenile cuckoo, on its first perilous migration, the adults long since departed from Britain in June or July. The bird alighted just out of view in an ash tree, expertly judging my eye-line as so many cuckoos had when I was studying them on Dartmoor. One more glimpse, then it was gone. In a single moment: the magic of low-carbon birding, and a reminder of what is at stake.

Climate Change in the Kalahari

Amanda Bourne

The Kuruman River Reserve, a beautiful slice of summer-rainfall savannah in the southern Kalahari, is characterised by undulating red sand dunes dotted with iconic camelthorn trees and teeming with wildlife. It is hot, with daily summer maximum air temperatures averaging 35°C, and dry, with total annual rainfall averaging 185 millimetres, a seventh of the annual average rainfall in the UK. High temperature extremes and dry summers have become more common and more severe in recent years, making this a good place to study the impacts of high temperatures and drought on birds.

I first came here as a PhD student, as part of the Hot Birds Research Project, which is part of a wider international effort to study the effects of high temperatures and drought on arid-zone species in South Africa and beyond, including southern yellow-billed hornbills, jacky winters, common fiscals, white-browed sparrow-weavers, red larks and fork-tailed drongos.

I joined the Hot Birds Research Project to study the effects of high temperatures and drought on behaviour, reproduction and survival in a cooperatively breeding passerine endemic to the arid savannahs of the Kalahari: the southern pied babbler (hereafter 'pied babbler'). Pied babblers are slightly smaller than song thrushes and the males and females look the same. They are highly social, living in territorial groups ranging in size from 3 to 15 adults. Each group consists of a single breeding pair with subordinate helpers that are usually, but not always, the offspring of the breeding pair. Dominant pairs monopolise almost all breeding activity, and subordinates rarely breed even when unrelated potential breeding partners are present in the group. Pied babblers breed in open-cup nests during the austral summer, typically between September and April, when it is hottest.

Associate Professor Amanda Ridley has studied pied babblers at the Kuruman River Reserve since 2003, and for my research I used a combination of her long-term life history dataset and the data collected during three years of my own fieldwork.

In order to understand how pied babblers respond to high temperatures and drought in the wild it is important to be able to monitor individuals in a group. For this I benefitted from the existence of a population of pied babblers that is comfortable with the presence of researchers. Each bird had a unique combination of colour rings on its legs and I knew them by name (Como, Bimbo, Winter, Misty …). These birds were used to having people observe them from just a couple of metres away and would hop onto a scale to be weighed in exchange for a bit of crumbled egg yolk or a mealworm. I was also able to follow the birds as they went about their business, observing their behaviour in their natural environment as they responded to naturally occurring extreme weather conditions in real time.

So what did I learn? I learned that hot weather increased the risk of mortality during early development in pied babblers.[1] Pied babbler eggs were half as likely to hatch when average daily maximum temperatures exceeded 35.5°C during incubation. Nestlings were less likely to fledge when temperatures exceeded 37.3°C between hatching and fledging. And dependent fledglings were less likely to survive to the point at which they can feed themselves (about 90 days after fledging) when temperatures exceeded 36.6°C between fledging and independence.[2] In fact, when average daily maximum temperatures exceeded 38°C, I did not observe a single successful breeding attempt across 15 years of data.[3] Temperatures in the mid- to high 30s are not unusual in the Kalahari and they are becoming more common; the frequency, duration and severity of extreme heat waves are growing with climate change.

What exactly was causing breeding failure at high temperatures? I measured various factors related to physical condition, physiology, behaviour and survival in pied babblers at different life stages (adults, juveniles, nestlings) and correlated these measurements with temperature and rainfall. Incubating birds lost weight and showed signs of becoming dehydrated when incubating for long periods on hot days. Usually, pied babblers take turns to incubate their eggs and attend nests continuously throughout the day. On very hot days, however, I noticed that the birds would leave their nests unattended more often and for longer periods of time, eventually abandoning their nests entirely.[4] Reduced nest attendance may have exposed the eggs to lethally high temperatures in the nest, causing them to become unviable, or put them at increased risk of predation.

High temperatures also affect birds in other ways. I measured nestling body mass, wing length and leg length in the morning and evening on days 5 and 11 after hatching.[5] Nestlings grew less vigorously during the day when the weather was hot than during cooler weather. This pattern was evident during both the fast-growing stage around 5 days after hatching and the slower-growing phase closer to fledging age, around 11 days after hatching. High temperatures directly affected nestling growth and influenced parental care behaviours, resulting in reduced provisioning to nestlings. Together, these effects led to reduced fledging success.

High temperatures did not only influence breeding: I also found that high temperatures in the first couple of months after fledging reduced the chances of juvenile pied babblers being recruited into the adult population, and that high temperatures during the breeding season reduced survival of breeding adults from one year to the next.[6] In short, high temperatures were the dominant factor influencing reproductive failure and mortality during all the life stages I studied. The effects of high temperatures on breeding and on survival from one year to the next were even more severe when high temperatures occurred in combination with drought.

With temperatures increasing rapidly in the Kalahari and hot droughts predicted to become more frequent and severe, pied babblers will increasingly experience conditions that inhibit both successful breeding and the survival of two important age groups: juveniles being recruited into the adult population and experienced breeding adults. Impacts of hot droughts on the survival rates of experienced breeding adults are particularly concerning as it is these birds that contribute most to population recovery after extreme events.[7] Pied babblers are relatively long-lived and well adapted to the natural variability of their arid environment, but an increase in the frequency and severity of hot weather extremes, especially hot droughts, could undermine population recovery during good years, and population growth and persistence overall, leading ultimately to localised extinctions for this species.

Although acute, lethal effects of high temperatures and drought as a direct result of dehydration, starvation and hyperthermia do occur in wild populations of pied babblers, they are much less common than the chronic, sub-lethal effects of high temperatures and drought acting on individuals via effects on body condition, reproduction, foraging efficiency, survival from one year to the next and, ultimately, population persistence that I have described here. My research found strong evidence for chronic and severe sub-lethal effects of high temperatures and drought on reproduction and survival from one year to the next in pied babblers.

Many other researchers in Southern Hemisphere environments are reporting similar effects of high temperatures and drought on reproduction and survival in desert birds. These are worrying findings. Climate change is advancing at an extremely rapid rate and adverse conditions are likely to result in population declines due to widespread breeding failure and subsequent low replacement and recruitment in a range of different species, even those that appear well adapted to hot and dry conditions. We should not underestimate the extent to which climate change will affect our beloved bird communities, nor should we underestimate the challenge of avoiding a frightening climate future.

It is highly likely that future environmental conditions will be more dangerous than most people currently believe. The scale of the threat to ourselves and all other living things is so great that it can even be difficult for well-informed experts to grasp. An effective international response has yet to emerge, but a realistic appreciation of the scale of the challenge we face and the kinds of actions we can take might allow us to chart a less-ravaged future. We can all contribute to this future by making informed decisions about the ways in which we enjoy birdwatching and by talking with joy and enthusiasm about our low-carbon birding experiences.

30

Unsettling Journeys

Kieran Lawrence

Sadly, and despite the best efforts of conservationists, the populations of many birds continue to be threatened globally and are declining due to many factors: agricultural intensification affecting farmland species, persecution of raptors in British uplands, illegal hunting in the Mediterranean, invasive species killing seabirds in remote oceanic islands... Presenting potentially the greatest danger is, however, climate change. This is the focus of my research as a doctoral student at Durham University; I assess the past impacts of climate change on migratory birds and predict the potential impacts of further change.

Climate change has multiple negative impacts for many species, not just those that migrate. Firstly, many temperate species are suffering from phenological mismatch, that is the desynchronisation of the timing of events in an individual's annual cycle and that of the ecosystem in which they dwell. This is a result of warmer temperatures leading to advancements in the timing of spring each year, with which bird populations are unable to keep up. For example, oak tree bud burst and 'leaf out' is occurring on average earlier each year, due to warmer temperatures. In response, oak processionary caterpillars hatch out earlier to ensure that they have the greatest supply of newly emerged leaves to feed on. Unfortunately, blue tits have been less capable of advancing their breeding cycle and, therefore, when peak food demand from chicks occurs, there are fewer caterpillars. This may lead to the reduced survival of chicks to fledging and contribute to population declines.

Another potential impact is that of shifts and losses of species' distributions. Higher temperatures result in the movement of species' ranges towards the poles, as they try to track the cooler temperatures to which they are adapted. For many species, however, this results in the overall reduction of their distribution range, as they are unable to keep pace with these shifts. While southerly parts of a species' range become too hot to occupy, more northerly areas become hospitable but are too far away for individuals to colonise. This is especially true for those species that do not usually disperse far from the sites at which they hatched and, as such, do not explore much of their surroundings for suitable habitat. Additionally, species that are distributed at the poleward edge of continents, for example sanderling breeding on the Arctic tundra or African penguins on the sandy beaches of Southern Africa, may be squeezed against the edges of landmasses, until eventually they have nowhere to go.

Finally, climatic changes may affect interactions between species. For instance, warmer winters are linked to the decline in sand eel populations in the North Atlantic, which is likely to have knock-on effects on the productivity of seabirds, which rely heavily on these populations as a food source. Additionally, certain species can adapt more rapidly to climatic changes and, therefore, are able to outcompete others. The latter is especially prevalent in scenarios involving invasive species, which may take advantage of more favourable winters, longer growing seasons and the lack of natural parasites or predators.

While resident populations are affected by these changes, for migratory species they bring additional dangers. The use of multiple areas across their annual cycle, breeding grounds, stopover sites and non-breeding grounds renders migratory species more susceptible to environmental change as there are simply more locations within which to be impacted. Probably as a result, migratory species have declined far more rapidly than their resident counterparts within the last 50 years. Firstly, phenological mismatch is more severe for migrants than resident species, such as in pied flycatchers compared with blue tits, both hole-nesting woodland species. This is because migrants must judge conditions in breeding grounds while still in nonbreeding areas, often in another hemisphere. This has led to the delayed arrival of individuals, in terms of the timings of spring events in the ecosystem on the breeding grounds. This in turn has led to an even greater gulf between the peak food demand of chicks during breeding and prey emergence.

Furthermore, the breeding and nonbreeding ranges of long-distance migrants, specifically, are often located in opposite hemispheres. As such, poleward shifts result in these ranges moving in opposite directions, leading to an increase in the distance these species are required to migrate each year. Migration is a period of particularly high mortality for birds due to unknown distribution of food resources and predators. Therefore, any increase in migration distance or any additional stopovers required for refuelling may further contribute to population declines. What is more, these changes are likely to increase the duration of these journeys, leading to later arrivals at the breeding grounds and exacerbation of the impacts of phenological mismatch.

Finally, climate change has resulted in more favourable winter conditions in higher latitudes; however, migratory species do not benefit as they are in the tropics during this period. Instead, we see increased overwinter survival in species which are resident at these higher latitudes, which are then in better condition for the breeding season. As a result, they can outcompete migrants for territories and nesting locations, and this has been observed for blue tits and pied flycatchers.

Climate change has had and will continue to have a huge impact on birds globally, particularly on migratory species. Yet these impacts are difficult to see unless you look closely and over a large timescale. This invisibility makes the issue, for now at least, easy to ignore. But this does not alter the fact that humans are having a deleterious effect on the natural world. I acknowledge that this issue is much bigger than any one individual and that change needs to happen in governments and big corporations, but surely that should not stop us from doing what we can? Cutting down on your use of high-emission travel is the biggest single contribution you can make in your day-to-day life. More specifically, attitudes to flying need to change, not least because the decision is as simple as not getting on a plane. Hour for hour, the quickest way a person can warm the planet is by flying on a commercial aircraft.

I do understand why people who are able to take multiple long-haul flights a year to see birds do so. Despite being only 24 years of age, I have done my fair share of travelling—across the UK, Europe, Africa and North America. I am incredibly grateful for the opportunities I have had to visit some of the most bird-rich locations in the world (*shout-out to you, Dad*). Birding trips, and particularly those abroad, have given me some of my greatest experiences. However, I have come to realise that I and everyone else need to adjust our behaviour—not least because of the findings of my research.

I want to make it clear that I am not suggesting that people need immediately to go cold turkey. For many, holidays are an escape from normal life and can benefit mental health tremendously. However, with more than 50% of the UK's population and 90% of all people across the globe coping without taking a single flight, can those that do fly not join them, and at the same time find pleasure in the wealth of nature nearer to our homes? One may argue that many of the places that birders visit would not be visited for any other reason. The money generated by this tourism is invaluable for many local communities and the conservation projects they run. The year 2020 highlighted this more than any other year; with birders being unable to travel due to the Covid-19 pandemic, many local conservation projects have been hit hard. Asa Wright Nature Centre and Lodge, located in Trinidad, a household name in world birding, had to close its doors for good, apparently unable to stay afloat with the lack of custom. It is, however, potentially more important to question why these places have become so heavily reliant on ecotourism and remember that the money gained through these avenues may well be outweighed by the income which will be lost if the climate continues to change in these places. Regardless, we need to look at our current habits and assess whether we think they are justifiable. For starters, are multiple holidays a year necessary, or would one suffice? Are holidays abroad even necessary at all?

But rather than just focusing on what needs to be cut out, let us also remember what can be gained from staying closer to home or altering our travel habits. I have recently realised how many locations in the UK I have yet to explore. I could quite easily spend the rest of my life exploring my own country and still not have seen all its wildlife by the end of my days. Many temperate ecosystems are also fortunate enough to play host to mass migration events, the flavour of birding seemingly changing month to month. However, if this is still not enough to satisfy you and you need to take a break from your native flora and fauna, can you travel more sustainably? Rail transport, especially outside of the UK, is becoming incredibly quick and efficient. Many routes can even be quicker and cheaper than flights, when you take into account wait times in airports and nights in hotels at either end. Rail travel is typically far more comfortable than a budget airline and lets you experience a lot more on your journey, even enabling birding from a window along the way. This is something I have explored within the last year and I can see that most of my birding holidays will be by means of train travel in the future—my next trip, from Newcastle to Hungary, is already planned. And if supporting local tourism is what is driving some of your flying to holiday in distant places, might you be able to donate without having to visit? You can then rest easy in the knowledge that, not only are you aiding the conservation of some of the world's most threatened and charismatic birds, but you have done it without contributing to climate change.

I will finish with some food for thought. For many birders who have in the past taken frequent flying for granted, flying less is likely to be a difficult and completely lifestyle-altering change. However, we need to remember how important it truly is. You might enjoy your next trip to Texas or Bulgaria or the Yellow Sea, but would this enjoyment be lessened by the knowledge that those populations of golden-cheeked warbler, red-breasted goose and spoon-billed sandpiper might not exist by the start of the next century?

31

Witness to Extinction

Alexander Lees

It is late September 2004, and I am marooned on a forest island amid a sea of pasture in sweltering tropical heat. A herd of white Brahman cattle have encircled the small patch of Amazonian rainforest near the town of Alta Floresta in Mato Grosso state that I am surveying for my PhD. It feels like a siege—one that the cattle are winning as pasture replaces rainforest across millions of hectares of the basin. This particular forest patch is largely devoid of birdlife: too small to support most forest-associated species and ravaged by 'edge effects', such as drying winds, which alter the forest micro-climate. A single pair of black-throated antbirds are the only representatives of the locally super species-rich family to have succeeded in persisting in this four-hectare fragment.[1] Leaving the forest, visibility is limited—ash is falling from the sky as the whole of the region is cloaked in smoke from thousands of fire foci—this is the 2004 peak in Amazonian deforestation and my second year of work in the basin, a place that is home to the greatest expression of avian biodiversity found anywhere on earth and one of the most imperilled.

I have gone on to dedicate most of my adult life to trying to understand the impacts of forest loss, fragmentation and degradation on Amazonian ecosystems. After finishing my PhD in 2008, I spent a couple of in-limbo years in consultancy, mostly doing bird surveys to assess impacts from proposed offshore wind farms, before moving back out to Brazil to work as a postdoctoral fellow at the Goeldi Museum, based in Belém, the capital of the Brazilian state of Pará. I ended up leading the bird work on what would become one of the largest ecosystem accounting exercises ever undertaken as we investigated how levels of biodiversity, economic development and ecosystem services covaried across a gradient of forest loss, under the auspices of the Sustainable Amazon Network.[2]

By 2009, deforestation rates had fallen sharply, the product of government controls, international pressure and the vagaries of economics.[3] Things were looking rosier for the largest remaining tropical rainforest on the planet. But then 2010 dealt a major El Niño event, one supercharged by climate change in a warming Amazon, triggering an intense regional drought and associated wildfires which are estimated to have released 2.2 billion tonnes of carbon following tree mortality and reduced carbon assimilation and tree growth.[4] Even though there was less felling, fires spreading through a drought-stressed forest whose resilience had been weakened by widespread illegal logging wreaked catastrophic Amazonian biodiversity losses. These impacts resulting from forest disturbance (reduction of habitat quality) covered a far greater area than that impacted by deforestation (total habitat loss).[5] In some areas, even undisturbed forest burned, something that scientists had not expected to happen as, unperturbed by chainsaws, these forests can usually retain enough humidity to repel any flames escaping from neighbouring human-modified landscapes—where fire is used as an agricultural tool. A tipping point had evidently been reached.

The 2015–16 El Niño event was even stronger than that of 2010, and a vast area of the central Amazon in the Santarém region of Pará state burned, including many of the permanent forest plots we had established as part of our research network. This gave us a unique opportunity to understand the divergent impacts of different stressors on Amazonian wildlife—the combination of climate change and degradation—which in synergy act as magnifiers in the biodiversity crisis.[6] By this point, evidence of the impacts of climate change on untouched forests was beginning to emerge. Anecdotes were the first indications: birders who had visited sites like Alta Floresta had been reporting declines in species abundance and the virtual disappearance of some species—especially the terrestrial insectivorous antbirds, antpittas and antthrushes. These species need a humid forest understorey and, in much the same way as they disappear from small forest fragments as the microclimate changes, now these species were winking out from the edge of their range from *within* large tracts of undisturbed forest. For example, I had encountered the striated antthrush of the eastern subspecies *fulvipectus* on several occasions during my PhD—a taxon that may well be elevated to species status in the future—which as a species is likely to be straight onto the global Red List of threatened species. Its voice has now largely been silenced from Alta Floresta and is increasingly becoming the stuff of legend among birders. A recent paper from the heavily inventoried Biological Dynamics of Forest Fragments Project, north of Manaus in the Brazilian Amazon, documented population crashes in a *whole suite* of understorey species.[7] Climate change is strongly suspected to be the primary driver of the declines as little else has changed in these remoter northern Amazonian forests.

Forest loss and fires grabbed global headlines in the summer of 2019 as a change in political leadership to one characterised by rampant anti-environmental views from within government effectively gave the green light to deforestation—rates that had previously been declining steadily began rising again. It was smoke from pyres of felled trees eventually drifting thousands of kilometres south-east to occlude the sun over the metropolis of São Paulo that catapulted the story onto the front pages of the global press. For those of us working on Amazonian conservation issues it was both a hectic and heart-breaking time as the global media scrambled for 'experts' to explain what was happening. Articles we wrote for outlets like *The Conversation* accrued vast readerships and all sorts of promises were made by governments to combat the crisis.[8] By August 2020, however, the world had been engulfed in the Covid-19 pandemic and the inevitable repeat of the 2019 fires crisis that followed in 2020 scarcely managed to break into print or social media, let alone accrue front page space. People had other things to be worried about. But the problem has not gone away as the collective attention has drifted, as we wrote in 2018:

> Global warming is not proceeding at the same rate across the planet. Although the greatest absolute temperature increases are occurring at higher latitudes, the tropics are already some of the hottest places on the planet and have the lowest inter-annual temperature variability. Consequently, they will be the first areas to experience significantly warmer climates than the present day and will endure climatic conditions without present-day equivalents.[9]

There exists a portfolio of solutions available to us to stop the loss, fragmentation and degradation of Amazonian forests by leveraging political and economic interests. Chief among these are policy-related controls (moratoria) of exports of products tainted by recent deforestation, especially beef and soybeans. Cognitive dissonance is, however, a peculiar beast—I used to eat lots of cheap, tasty organic beef while doing fieldwork and only stopped doing so as late as 2014, followed shortly after by other meat products. This was despite my in-depth scientific knowledge of the problems that this diet caused, many of which I had seen first-hand. As a driver of climate and land-use change, two processes which are intimately linked, it is second to none. Animals can be farmed to deliver biodiversity co-benefits via grazing for rewilding and restoration, but such instances are vanishingly rare when compared with the status quo. Shifts in diet will be a key means to averting biodiversity loss and dangerous climate change.

Asking an ecologist about travel carbon budgets is increasingly a source of embarrassment: we travel inter-continentally for fieldwork, conferences, teaching and so on. Slowly, that is changing where possible. The Covid-19 pandemic forced us to move project meetings online, and also conferences, with some degree of success; there are major challenges ahead, but the direction of travel is reduction. On a personal level I have not been on a foreign birding holiday since 2017. Yet in the first seven months of 2019 I flew to Brazil to give a keynote speech at a conference, flew to Cyprus to examine a PhD and flew to both Portugal and Tanzania to teach students on field courses. The trip to Brazil in July 2019 was the last time I left England (as of writing in October 2021). The PhD viva in Cyprus could have been easily switched and I have done several online since, which seems likely to be the future of such exams. It is a shame to lose that personal contact but there are mighty trade-offs at stake.

I am fortunate that birding has often been a job rather than just a hobby. A hobby which has been very high carbon in the past, but since returning from the USA where I worked as a postdoc after five years in Brazil, I have dramatically reduced my travel for pleasure. In my first autumn back in England, living in the Derbyshire Peak District (as an academic based at Manchester Metropolitan University), my wife and I made regular weekend trips to both the east and west coasts by car. In an exceptional autumn for eastern vagrants we found a red-flanked bluetail in Lincolnshire, but just as exciting was bumping into two yellow-browed warblers—one in the middle of Manchester city centre and another at a tiny Peak District reservoir in Derbyshire. Over the following years my horizons collapsed somewhat, and most birding became very local, with aspirations adjusted accordingly. Rarity is always relative. Prior to leaving for Brazil I had lived in Norfolk for over a decade and most trips to the north Norfolk coast were motivated by the desire to find nationally rare birds—with some expectation of success. The background species, including rarer breeding and migrant species, were typically also-rans to which I paid scant attention—many such species I have scarcely seen for a decade but still have hopes of one day finding near my home in the valley of Longdendale. I look back at that time in Norfolk and can honestly say that most days tended to end in frustration because goals were set so high; my levels of happiness after a day in the field are as high, if not higher, now, in an area where the chances of finding national rarities is orders of magnitude lower.

Part of this happiness has been derived from an increasing addiction to the eBird platform (of which I am both an active scientific consumer and amateur contributor of data),[10] which has made me pay far more attention to the ebb and flow of commoner species through the seasons and created a compulsion to complete the weekly bar charts across all my local birding sites and provide rich media (photos and sound recordings) for each species. This obsessive-compulsive patching has made birding more rewarding, and more hours in the field equate to greater success. For instance, during the second Coronavirus wave in late winter 2021, I walked around Bottoms Reservoir (five minutes on foot from my door) almost every morning, and was rewarded with local scarcities (think deep oligotrophic, birdless reservoir that makes Bartley Reservoir look like Welney) like single, grounded black-legged kittiwake, white-fronted goose and common scoter, eclipsed by four black terns and two incongruous gannets in the autumn. Away from the Peak District we have visited the Isles of Scilly for three family holidays; our travelling from Hadfield to Penzance by car generates 102 kilograms of CO_2 emissions for a petrol cost of £80. I would prefer to travel by train, 39.4 kilograms of CO_2, but the financial cost for the same journey by train for three of us is *six times* more expensive (and takes longer). We need to subsidise rail travel.

Beyond some modest success in finding rare birds off season on Scilly (citrine wagtail, woodchat shrike, European bee-eater, Wilson's storm-petrel, spotted crake on the last three trips in April and August), I spent most of last autumn trying to find Siberian vagrants and especially red-flanked bluetail on the moorland fringe at home. I failed, but proof of concept was provided by ringers on the opposite eastern flank of the Peak District who caught a bluetail on eBird's Global Big Day—for which I spent 14 hours and walked 24 kilometres in the field and on foot from home to no avail. Hope springs eternal.

Postscript: On the 29 November 2021, I found a Blyth's reed warbler in a snowstorm in Manchester city centre—never give up hope!

Afterword

The early 1990s were an exciting time to be a young bird-watcher concerned about nature. The environmental movement was growing rapidly, Greenpeace had a radical edge and words such as biodiversity and sustainable development sounded new. With the fall of the Iron Curtain, the planet was being treated as a single political arena and rainforests and colourful tropical frogs had come to symbolise a global nature. In my young, naïve eyes, it looked as if the world was finally getting ready for action. The pinnacle of this growing international environmentalism was the 1992 Earth Summit in Rio de Janeiro, at that time the largest ever coming together of world leaders to discuss the environment. At that event, 12-year-old Severn Cullis-Suzuki reminded world leaders that 'you are what you do, not what you say' and asked them to make their actions reflect their words.

I read avidly about the Rio Summit and still keep newspaper clippings about it. But what evokes the most vivid memories of that time is Jonathon Porritt's book *Save the Earth*, a beautifully edited compilation of thematic essays, arresting images and short personal accounts by scientists, religious leaders, artists, authors and campaigners celebrating the fragile beauty of our planet and calling for an end to its destruction. I open the book today and find it difficult not to think of everything that has been lost, all the suffering that could have been avoided. Had the ecological and climate crises been taken seriously, the kind of incremental change being proposed today by mainstream politics would have sufficed. Instead, three decades of inaction have brought us dangerously close to climate tipping points where small changes could push parts of the Earth system into abrupt or irreversible change.

The fact that climate scientists find it difficult to sleep at night because they are worried about the future of their children, the fact that as I write these lines people are dying in India due to an unprecedented heatwave, reflects how dire the situation is. The scale of the challenge we face cannot be overstated. According to climate and sustainability scientists, nothing short of a global Marshall Plan-style transformation to a zero-carbon energy system and profound changes in behaviour will give us a chance of maintaining a liveable planet. It is clear by now that the time of waiting for politicians to lead is over. As Mike Clarke rightly argues in the Foreword to this book, and as research on social movements confirms, politicians do not lead, they follow: only 'when a sufficiently broad coalition of people and organizations mobilize around a common cause, reaching a critical "mass" and "momentum" for change … only then can politicians feel sufficiently emboldened to build their own coalitions that lead to government action'.[1]

Like many publications about the environment before and after 1992, Porritt's book ends with a sense of urgency and hope. I am ending this book not with a message of hope, but with a reminder that your actions matter more than you probably realise. In the same way that there are tipping points in the Earth system, there are tipping points in social and political life. Each of us can help in building this critical mass and momentum for change. It starts with individuals and groups in myriad localities and organisations creating coalitions for change around a shared agenda. It starts with new narratives about a better future that can bring together people from all walks of life—stories inspired not by ideology but by people's everyday experiences and the feeling of vulnerability that connects each of us to most of the people on Earth. And it grows by connecting these initiatives and ensuring that there is a strong ripple effect.

I see low-carbon birding as enabling this wider social transformation. At the most basic level, it convinces us of our own commitment and we begin to see ourselves as climate advocates. We learn to expect more from those with greater influence and power; we expand the horizon of our possibilities and ambition. Importantly, in our achievements and frustrations we learn to see more clearly how things could be different. Imagine, for example, if our neighbourhoods, villages and country roads had fewer cars and children could discover nature safely on bicycles. Imagine if everyone in your community was healthier because they spent more time walking, cycling and socialising rather than driving. Imagine if all the space used for car parking in our cities was turned into green spaces for wildlife and for growing fruit and vegetables. Imagine if, instead of building new roads, governments spent more money on improving and expanding walking, cycling and rail networks, and everyone—including poor, elderly and disabled people—could travel anywhere safely without a car. Imagine a countryside where fewer animals were killed on roads. Imagine if instead of a constant hum of traffic, our neighbourhoods and villages were quiet enough to notice, at any time of the day, the buzz of the bumblebee and the song of the blackbird.

And that is just the beginning. Imagine if our children, the future generations of birdwatchers, could look back at our time and say with respect and admiration that our actions reflected our words.

Javier Caletrío

Notes

Introduction

1 The calculations behind that figure are explained by Dr Hugh Hunt, Department of Engineering at Cambridge University, in this video: https://www.youtube.com/watch?v=rcYVLMtWcFk&feature=emb _logo

2 Chancel, L. (2021) *Climate Change and the Global Inequality of Carbon Emissions 1990–2020*. Paris: World Inequality Lab; Ghosh, E., Gore, T., Kartha, S., Kemp-Benedict, E. and Nazareth, A. (2021) *The Carbon Inequality Era: An assessment of the global distribution of consumption emissions among individuals from 1990 to 2015 and beyond*. Stockholm: Oxfam, Stockholm Environment Institute; Nielsen, K., Nicholas, K., Creutzig, F. and DietStern, P. (2021) The role of high-socioeconomic-status people in locking in or rapidly reducing energy-driven greenhouse gas emissions. *Nature Energy* 6: 1011–16.

3 On the need for collectively defined self-limitation see Kallis, G. (2019) *Limits: Why Malthus Was Wrong and Why Environmentalists Should Care*. Stanford, CA: Stanford University Press.

4 Gillespie, E (2020, 2 February) 7 reasons why I'm swapping flights for train travel in 2020. *Conde Nast Traveller*. https://www.cntraveller.com /gallery/reasons-to-travel-by-train

5 Guenther, G. (2020) Communicating the climate emergency: imagination, emotion, action. In C. Henry, J. Rockström and N. Stern (eds) *Standing up for a Sustainable World: Voices of Change*. Cheltenham: Edward Elgar Publishing, pp. 401–8.

6 Marsden, G., Anable, J., Chatterton, T., Docherty, I., Faulconbridge, J., Murray, L., Roby, H. and Shires, J. (2020) Studying disruptive events: innovations in behaviour, opportunities for lower carbon transport policy? *Transport Policy* 94: 89–101.

7 This pattern of change is illustrated in Jonathon Porritt's *The World We Made*. This excellent book is part of a burgeoning literature about environmental futures seeking to facilitate the identification of transition

pathways. The book is explicitly written to bring hope to a field—future studies—often associated with bleak perspectives, by showing that sustainability is not necessarily about the renunciation of earthly pleasures, lower quality of life or limiting one's cultural and geographical horizons, and that the experience, the knowledge and the technologies needed to begin making the world a more sustainable reality are already there. As compared to other studies, the book is of special interest for the way it conceives of a transition path as a sequence of interlinked, cascading events or 'shocks' (rather than simply the result of purposeful action) and the role of social and political tipping points in this unfolding of events. See Porritt, J. (2013) *The World We Made: Alex McKay's Story From 2050*. London: Phaidon. For a sociological discussion of such a transition pathway, see Urry, J. (2016) *What Is The Future?* Cambridge: Polity Press.

Chapter 1

1 Anderson, K. and Bows, A. (2011) Beyond 'dangerous' climate change: emission scenarios for a new world. *Philosophical Transactions of the Royal Society A* 369 (1934): 20–44.

2 Anderson, K. and Bows, A., Beyond 'dangerous' climate change: 20–44; Raupach, M. R., Davis, S. J., Peters, G. P., Andrew, M. R., Canadell, J. G., Ciais, P., Friedlingstein, P., Jotzo, F., van Vuuren, D. P. and Le Quéré, C. (2014) Sharing a quota on cumulative carbon emissions. *Nature Climate Change* 4: 873–9; Larkin, A., Kuriakose, J., Sharmina, M. and Anderson, K. (2017) What if negative emission technologies fail at scale? Implications of the Paris Agreement for big emitting nations. *Climate Policy* 18: 690–7.

3 Bows-Larkin, A. (2015) All adrift: aviation, shipping, and climate change policy. *Climate Policy* 15: 681–702.

4 Balmford, A., Cole, L., Sandbrook, C., and Fisher, B. (2017) The environmental footprints of conservationists, economists and medics compared. *Biological Conservation* 214: 260–9.

5 Heinrich-Böll-Stiftung / Airbus Group (2016) Oben – Ihr Flugbegleiter. Berlin: Heinrich-Böll-Stiftung.

6 Holden, E. (2005) Attitudes and sustainable household consumption. The European Network of Housing Research (ENHR) International Housing Conference, Reykjavik, Iceland, 29 June–3 July; Barr, S., Shaw, G. and Coles, T. E. (2011) Times for (un)sustainability? Challenges and opportunities for developing behavior change policy: a case-study of consumers at home and away. *Global Environmental Change* 21: 1234–44.

7   Hirst, A. (2017) SK58 Birders: 25 years of watching a 10-km square. *British Birds* 110: 367–424.

8   https://www.patchworkchallenge.com/

9   Chancel L. and Piketty, T. (2015) *Carbon and Inequality: From Kyoto to Paris*. Paris: Paris School of Economics; Gore, T. (2015) *Extreme Carbon Inequality: Why the Paris Climate Deal Must Put the Poorest, Lowest Emitting and Most Vulnerable People First*. Nairobi: Oxfam. https://policy-practice.oxfam.org.uk/publications/extreme-carbon -inequality-why-the-paris-climate-deal-must-put-the-poorest-lowes -582545

10 Gössling, S., Ceron, J. P., Dubois, G., and Hall, M. C. (2012) Hypermobile travellers. In S. Gössling, S. and P. Upham (eds), *Climate Change and Aviation: Issues, Challenges and Solutions*. Oxford: Routledge, pp. 131–50.

11 Department for Transport (2014) *Public Experiences of and Attitudes towards Air Travel: 2014*. London: Department for Transport UK.

12 Piketty, T. (2014) *Capital in the Twenty-First Century*. Cambridge, MA: Harvard University Press.

13 The figure provided in the original article published in British Birds was '2% according to the aviation industry'. The figure of 4% is from Klöwer, M., Allen, M. R., Lee, D. S., Proud, S. R., Gallagher, L. and Skowron, A. (2021) Quantifying aviation's contribution to global warming. *Environmental Research Letters* 16: 104027.

14 Öko-Institut (2015) *Emission Reduction Targets for International Aviation and Shipping*. Brussels: European Parliament.

15 Evans, S. (16 October 2016) Analysis: Aviation to consume half of UK's 1.5C carbon budget by 2050. *Carbon Brief*. https://www .carbonbrief.org/analysis-aviation-to-consume-half-uk-1point5c -carbon-budget-2050

16 The figure provided in the original article published in British Birds was '5% of the world's population'. Different studies give different numbers usually oscillating between less than 5% and 20%. According to Stefan Gössling and Andreas Humpe, only 2% to 4% of global population flew internationally in 2018; see Gössling, S. and Humpe, A. (2020) The global scale, distribution and growth of aviation: Implications for climate change. *Global Environmental Change* 65: 102194.

17 Source Centre for Aviation website: https://centreforaviation.com/

18 Bows-Larkin, A., Mander, S., Traut, M. B., Anderson, K. and Wood, R. (2016) Aviation and Climate Change – the continuing challenge. In Blockley, R. H. and Shyy, W. (eds), *Encyclopedia of Aerospace Engineering*. Oxford: Blackwell.

19 Anderson, K. (1 January 2014) Is flying still beyond the pale? *New Internationalist*. https://newint.org/sections/argument/2014/01/01/flying-still-beyond-the-pale/

Chapter 2

1 Some commentators argue that the focus should be, for instance, on the 100 oil and coal companies responsible for 71% of emissions. This means focusing on where the supply chain starts. If we look at where it ends we find the people who consume the final products from fossil fuels. Prominent among these are the richest 10% who through their direct and indirect energy consumption are responsible for half of all global emissions.

2 Nielsen, K. S., Nicholas, K. A., Creutzig, F. *et al.* (2021) The role of high-socioeconomic-status people in locking in or rapidly reducing energy-driven greenhouse gas emissions. *Nature Energy* 6: 1011–16.

3 Newell, P., Daley, F. and Twena, M. (2021) *Changing Our Ways? Behaviour Change and the Climate Crisis.* Cambridge: Cambridge Sustainability Commission on Scaling Behaviour Change. https://www.rapidtransition.org/wp-content/uploads/2021/04/Cambridge-Sustainability-Commission-on-Scaling-behaviour-change-report.pdf

4 Gächter, S. and Renner, E. (2018) Leaders as role models and 'belief managers' in social dilemmas. *Journal of Economic Behavior & Organization* 154: 321–34.

5 Westlake, S. (2017) A counter-narrative to carbon supremacy: Do leaders who give up flying because of climate change influence the attitudes and behaviour of others? Unpublished MA thesis, Cardiff University.

6 This argument is clearly explained in Kimberly Nicholas's recent book. See Nicholas, K. (2021) *Under the Sky We Make: How to Be Human in a Warming World.* New York: G. P. Putnam's Sons. See also the chapter entitled 'The personal is political: How to be a good climate citizen' in Rebecca Willis's book *Too Hot to Handle.* Willis, R. (2020) *Too Hot to Handle: The Democratic Challenge of Climate Change.* Bristol: Bristol University Press. Atmospheric scientist Peter Kalmus has summarised the argument neatly: 'Collective action enables individual action (by shifting systems) and individual action enables collective action (by shifting cultural norms). Visible, conspicuous individual action is also collective action. We will not get a carbon fee and dividend, for example, until the grassroots care enough about climate change.'

The quote is from Kalmus's acceptance speech for the Transition US Walking the Talk Award.

7 United Nations Environment Programme (2020) *Emissions Gap Report 2020*. Nairobi: UNEP.

8 Gore, T. (2021) *Carbon Inequality in 2030. Per Capita Consumption Emissions and the 1.5°C Goal*. London and Oxford: Institute for European Environmental Policy and Oxfam.

9 Oswald, Y., Owen, A. and Steinberger, J. (2020) Large inequality in international and intranational energy footprints between income groups and across consumption categories. *Nature Energy* 5: 231–9.

10 Ivanova, D. and Wood, R. (2020) The unequal distribution of household carbon footprints in Europe and its link to sustainability. *Global Sustainability* 3: e18 1–12.

11 Stoddard, I., Anderson, K., Capstick, S., Carton, W. *et al.* (2021) Three decades of climate mitigation: why haven't we bent the global emissions curve? *Annual Review of Environment and Resources* 46: 653–89.

12 Department for Transport (2021) *Transport and Environment Statistics: 2021 Annual Report*. London: Department for Transport. https://assets.publishing.service.gov.uk/government/uploads/system/uploads/attachment_data/file/984685/transport-and-environment -statistics-2021.pdf

13 Brand, C., Anable, J. and Morton, C. (2018) Lifestyle, efficiency and limits: modelling transport energy and emissions using a socio-technical approach. *Energy Efficiency* 12: 187–207.

14 Anderson, K., Broderick, J. F. and Stoddard, I. (2020) A factor of two: how the mitigation plans of 'climate progressive' nations fall far short of Paris-compliant pathways. *Climate Policy* 20: 1290–304.

15 Anderson, K. (2019) Aligning UK car emissions with the Paris Agreement. Provisional budget analysis for DecarboN8. https://decarbon8.org.uk/aligning-uk-car-emissions-with-the-paris -agreement/. See also Milovanoff, A., Posen, I. D. and MacLean, H. L. (2020) Electrification of light-duty vehicle fleet alone will not meet mitigation targets. *Nature Climate Change* 10: 1102–7.

16 The annual mileage per car in the UK in 2019 was 8,700 miles. The annual per capita mobility by car in England in the period 2015–2020 was around 6,600 miles driven during 780 trips. Of this only 19 trips were long-distance travel (over 50 miles) but these represent 30% of total mileage (1,900 miles). See Department for Transport. (2020). *National Travel Survey, 2002–2019*, 6th edn. UK Data Service.

17 Heuwieser, M. (2017) *The Illusion of Green Flying*. Vienna: Finance &
    Trade Watch. https://stay-grounded.org/wp-content/uploads/2019/02
    /The-Illusion-of-Green-Flying.pdf
18 Graver, B., Rutherford, D. and Zheng, S. (2020) *CO_2 Emissions
    from Commercial Aviation 2013, 2018 and 2019*. Washington:
    International Council on Clean Transportation. https://theicct.org/
    sites/default/files/publications/CO2-commercial-aviation-oct2020.pdf
19 Information provided by the International Council on Clean
    Transportation using data from two reports published in 2020 by
    the International Energy Agency and the International Air Transport
    Association. See International Energy Agency (2020) *Energy
    Technology Perspectives 2020*. Paris: IEA. https://iea.blob.core.windows
    .net/assets/7f8aed40-89af-4348-be19-c8a67df0b9ea/Energy_
    Technology_Perspectives_2020_PDF.pdf; International Air Transport
    Association. (2020) *Economic Performance of the Airline Industry*.
    Montreal: IATA. https://www.iata.org/en/iata-repository/publications
    /economic-reports/airline-industry-economic-performance-june-2020
    -report/
20 Hof, C., Voskamp, A., Biber, M., Böhning-Gaese, K., Engelhardt,
    E., Niamir, A., Willis, S. and Hickler, T. (2018) Bioenergy cropland
    expansion may offset positive effects of climate change mitigation for
    global vertebrate diversity. *PNAS* 115: 13294–5; Muscat, A., de Olde,
    E. M., de Boer, I. J. M. and R. Ripoll-Bosch, R. (2020) The battle for
    biomass: a systematic review of food-feed-fuel competition. *Global
    Food Security* 25: 100330.
21 Transport & Environment (2018) *Roadmap to Decarbonising European
    Aviation*. Brussels: Transport & Environment. https://www.transpo
    rtenvironment.org/wp-content/uploads/2021/07/2018_10_Aviation
    _decarbonisation_paper_final.pdf
22 Fellow Travellers (2018) Electric dreams: the carbon mitigation
    potential of electric aviation in the UK air travel market. https://s3
    -eu-west-1.amazonaws.com/media.afreeride.org/documents/Electric
    +Dreams.pdf
23 Anderson, K. and Peters, G. (2016) The trouble with negative
    emissions. *Science* 354: 182–3.
24 On frequent flyers levy, see Chapman, A., Murray, L., Carpenter,
    G., Heisse, C. and Prieg, L. (2021) *A Frequent Flyers Levy: Sharing
    Aviation's Carbon Budget in a Net Zero World*. London: New
    Economics Foundation. https://neweconomics.org/uploads/files/
    frequent-flyer-levy.pdf (17 August 2021). On changing travel norms,
    see Capstick, S., Khosla, R. and Wang, S. (2020) Bridging the gap:

The role of equitable low-carbon lifestyles. In *Emissions Gap Report 2020*. United Nations Environment Programme. Nairobi. https://wedocs.unep.org/handle/20.500.11822/34432

25 Akenji, L., Bengtsson, M., Toivio, V., Lettenmeier, M., Fawcett, T., Parag, Y., Saheb, Y., Coote, A., Spangenberg, J. H., Capstick, S., Gore, T., Coscieme, L., Wackernagel, M. and Kenner, D. (2021) *1.5-Degree Lifestyles: Towards A Fair Consumption Space for All.* Berlin: Hot or Cool Institute. https://hotorcool.org/wp-content/uploads/2021/10/Hot_or_Cool_1_5_lifestyles_FULL_REPORT_AND_ANNEX_B.pdf

26 Funk, C. (2021) *Drought, Flood, Fire: How Climate Change Contributes to Catastrophes.* Cambridge: Cambridge University Press. On the problems of forest preservation offsets see the articles by investigative journalist Lisa Song: Song, L. (22 May 2019) An even more inconvenient truth: why carbon credits for forest preservation may be worse than nothing. *Propublica.* https://features.propublica.org/brazil-carbon-offsets/inconvenient-truth-carbon-credits-dont-work-deforestation-redd-acre-cambodia/; Song, L. and Moura, P. (26 August 2019) If carbon offsets require forests to stay standing, what happens when the Amazon is on fire? *Propublica.* https://www.propublica.org/article/if-carbon-offsets-require-forests-to-stay-standing-what-happens-when-the-amazon-is-on-fire; See also Skelton, A., *et al.* (11 December 2020) 10 myths about net zero targets and carbon offsetting, busted. *Climate Home News.* https://www.climatechangenews.com/2020/12/11/10-myths-net-zero-targets-carbon-offsetting-busted/

27 Waring, B. (2021, 23 April) There aren't enough trees in the world to offset society's carbon emissions – and there never will be. *The Conversation.* https://theconversation.com/there-arent-enough-trees-in-the-world-to-offset-societys-carbon-emissions-and-there-never-will-be-158181

28 EASAC (2018) Negative emission technologies: What role in meeting Paris Agreement targets? EASAC policy report 35. https://easac.eu/fileadmin/PDF_s/reports_statements/Negative_Carbon/EASAC_Report_on_Negative_Emission_Technologies.pdf

29 This argument is clearly developed in this article by three climate scientists: Dyke, J., Watson, R. and Knorr, W. (22 April 2021) Climate scientists: concept of net zero is a dangerous trap. *The Conversation.* https://theconversation.com/climate-scientists-concept-of-net-zero-is-a-dangerous-trap-157368

30 Fajardy, M., Köberle, A., MacDowell, N. and Fantuzzi, A. (2019) BECCS deployment: a reality check. Grantham Institute briefing paper 28. https://www.imperial.ac.uk/media/imperial-college/

grantham-institute/public/publications/briefing-papers/BECCS
-deployment---a-reality-check.pdf

31 Griscome, B. *et al.* (2017) Natural climate solutions. *PNAS* 114:
    11645–50.

32 See, for example, Corson, C. (2016) *Corridors of Power: The Politics of
    Environmental Aid to Madagascar.* New Haven, CT: Yale University
    Press.

33 There is an extensive and growing social science literature on wildlife
    conservation examining questions such as who controls local
    resources, whose values and interests inform conservation policies,
    who decides about tourism development, who benefits from tourism,
    how the revenue is distributed and how tourism affects wildlife. For
    an overview of some recent literature see Massarella, K. *et al.* (2021)
    Transformation beyond conservation: how critical social science
    can contribute to a radical new agenda in biodiversity conservation.
    Current Opinion in Environmental Sustainability 49: 79–87.

34 Brockington, D. and Igoe, J. (2006) Eviction for conservation: a
    global overview. *Conservation and Society* 4: 424–70.

35 See the Convivial Conservation initiative developed by Robert
    Fletcher, Bram Büscher and colleagues: https://convivialconservation
    .com/

36 Robert, R. and Büscher, B. (2020) Conservation basic income: a
    non-market mechanism to support convivial conservation. *Biological
    Conservation* 244: 108520.

37 Robert, R. and Büscher, B. (2020) *The Conservation Revolution:
    Radical Ideas for Saving Nature Beyond the Anthropocene.* London:
    Verso.

38 Mallapaty, S. (2021) The search for animals harbouring coronavirus—
    and why it matters. *Nature* 591: 2–8. https://doi.org/10.1038/d41586
    -021-00531-z. Reporting for the *Financial Times*, journalist Leslie
    Hook notes that 'There are about 1.6 million viruses on the planet in
    mammals and birds, of which about 700,000 could have the potential
    to infect humans. But of these, only about 250 have been identified
    in humans. The rest are still out there—they just haven't made the
    leap.' See Hook, L. (20 October 2020) The next pandemic: where is it
    coming from and how do we stop it? *Financial Times.* https://www.ft
    .com/content/2a80e4a2-7fb9-4e2c-9769-bc0d98382a5c

39 A good article on the limits (and injustices) of focusing solely on
    international tourists and the need to make parks in African countries
    more accessible to locals: Mzezewa, T. (2 September 2020) Do safari
    companies really want African travellers? *New York Times.* https://
    www.nytimes.com/2020/09/02/travel/Africa-safaris-local-tourism
    -coronavirus.html

Chapter 3

1 Campbell, O. J. and Moran, N. J. (2016) Phenology of spring landbird migration through Abu Dhabi island, United Arab Emirates, 2007–2014. *Sandgrouse* 38: 38–70.

2 https://patchworkchallenge.com/

Chapter 4

1 Stoker, G. (2018) Great Crested Grebe catching hirundines. *British Birds* 111: 402–5.

2 Brookes, R. (2008) House Martins eating elderberries. *British Birds* 101: 384.

3 Waldon, J. (2004) Sand Martins and House Martins nesting in a wall. *British Birds* 97: 352–3.

4 Newson, S. E., Moran, N. J., Musgrove, A. J., Pearce-Higgins, J. W., Gillings, S., Atkinson, P. W., Miller, R., Grantham, M. J. and Baillie, S. R. (2016) Long-term changes in the migration phenology of UK breeding birds detected by large-scale citizen science recording schemes. *Ibis* 158: 481–95; Woodward, I. D., Massimino, D., Hammond, M. J., Barber, L., Barimore, C., Harris, S. J., Leech, D. I., Noble, D. G., Walker, R. H., Baillie, S. R. and Robinson, R. A. (2020) *BirdTrends 2020: trends in numbers, breeding success and survival for UK breeding birds*. BTO Research Report 732. Thetford: British Trust for Ornithology. www.bto.org/birdtrends

5 Sparks, T., and Tryjanowski, P. (2007) Patterns of spring arrival dates differ in two hirundines. *Climate Research* 35: 159–64.

6 Woodward *et al.*, *BirdTrends 2020*.

7 Kettel, E. F., Woodward, I. D., Balmer, D. E. and Noble, D. G. (2020) Using citizen science to assess drivers of Common House Martin *Delichon urbicum* breeding performance. *Ibis* 163: 366–79.

8 Woodward *et al.*, *BirdTrends 2020*.

Chapter 15

1 www.audacityteam.org

2 'Nocmig' is the recording and monitoring of nocturnal migration and is the night-time counterpart to 'vismig', the monitoring of birds seen migrating in the day. As nocturnal migration is invisible to the naked eye, 'invismig' seemed appropriate.

Chapter 16

1 Birkhead, T. (2000) *Promiscuity*. London: Faber & Faber.
2 Schreur, G., Mayordomo, S. and Langlois, D. (14 September 2020) Fifty species in an hour: the remarkable mimicry of Common Redstart. *Birdguides*. https://www.birdguides.com/articles/ornithology/fifty-species-in-an-hour-the-remarkable-mimicry-of-common-redstart/
3 Brumm, H. (2004) The impact of environmental noise on song amplitude in a territorial bird. *Journal of Animal Ecology* 73: 434–40; Nemeth E., Pieretti N., Zollinger, S. A., Geberzahn N., Partecke J., Catarina M. and Brumm, H. (2013) Bird song and anthropogenic noise: vocal constraints may explain why birds sing higher-frequency songs in cities. *Proceedings of the Royal Society B: Biological Sciences* 280: 20122798; Slabbekoorn, H. and den Boer-Visser, A. (2006) Cities change the songs of birds. *Current Biology* 16: 2326–31.
4 Salaberria, C. and Gil, D. (2010) Increase in song frequency in response to urban noise in the great tit *Parus major* as shown by data from the Madrid (Spain) city noise map. *Ardeola* 57: 3–11.
5 Gil, D., Honarmand, M., Pascual, J., Pérez-Mena, E. and Macías Garcia, C. (2015) Birds living near airports advance their dawn chorus and reduce overlap with aircraft noise. *Behavioral Ecology* 26: 435–43.

Chapter 17

1 xeno-canto is an online global repository for recordings of bird songs and calls. https://xeno-canto.org/

Chapter 28

1 Thorup, K., Tøttrup, A. P., Willemoes, M., Klaassen, R. H. G., Strandberg, R., Vega, M. L., Dasari, H. P., Araújo, M. B., Wikelski, M. and Rahbek, C. (2017) Resource tracking within and across continents in long-distance bird migrants. *Science Advances* 3: 1–11.
2 Schneider, T., Bischoff, T. and Haug, G.H. (2014) Migrations and dynamics of the intertropical convergence zone. *Nature* 513: 45–53; Beresford, A. E., Sanderson, F. J., Donald, P. F., Burfield, I. J., Butler, A., Vickery, J. A. and Buchanan, G. M. (2018) Phenology and climate change in Africa and the decline of Afro-Palearctic migratory bird populations. *Remote Sensing in Ecology and Conservation* 5: 55–69; NASA (2000) The Intertropical Convergence Zone. *Earth Observatory*.

https://earthobservatory.nasa.gov/images/703/the-intertropical
-convergence-zone

3 Mills, L. J., Wilson, J. D., Lange, A., Moore, K., Henwood, B., Knipe, H., Chaput, D. L. and Tyler, C. R. (2020) Using molecular and crowd-sourcing methods to assess breeding ground diet of a migratory brood parasite of conservation concern. *Journal of Avian Biology* 51 (9). e02474

4 Beresford *et al.*, Phenology and climate change in Africa: 55–69.

5 Menzel, A., Sparks, T. H., Estrella, N., Koch, E., Aasa, A., Ahas, R., Alm-Kübler, K., Bissolli, P., Braslavská, O. G., Briede, A. and Chmielewski, F. (2006) European phenological response to climate change matches the warming pattern. *Global Change Biology* 12: 1969–76.

6 Saino, N., Rubolini, D., Lehikoinen, E., Sokolov, L. V, Bonisoli-Alquati, A., Ambrosini, R., Boncoraglio, G. and Møller, A. P. (2009) Climate change effects on migration phenology may mismatch brood parasitic cuckoos and their hosts. *Biology Letters* 5: 539–41.

Chapter 29

1 Bourne, A., Cunningham, S., Nupen, L., McKechnie, A. and Ridley, A. (2021) No sex-specific differences in the influence of high air temperatures during early development on nestling mass and fledgling survival in the Southern Pied Babbler (*Turdoides bicolor*). *Ibis* 164: 304–12.

2 Bourne, A., Cunningham, S., Spottiswoode, C. and Ridley, A. (2020) High temperatures drive offspring mortality in a cooperatively breeding bird. *Proceedings of the Royal Society B* 287: 20201140.

3 Moagi, L., Bourne, A., Cunningham, S., Jansen, R., Ngcamphalala, C., Ridley, A. and McKechnie, A. (2021) Hot days associated with short-term adrenocortical responses in a southern African arid-zone passerine bird. *Journal of Experimental Biology*, jeb.242535

4 Bourne, A., Ridley, A., McKechnie, A., Spottiswoode, C. and Cunningham, S. (2021) Dehydration risk is associated with reduced nest attendance and hatching success in a cooperatively breeding bird, the southern pied babbler *Turdoides bicolor*. *Conservation Physiology* 9: coab043.

5 Bourne, A., Ridley, A., Spottiswoode, C. and Cunningham, S. (2021) Direct and indirect effects of temperatures on fledging success in a cooperatively breeding bird. *Behavioral Ecology* 32.6: arab087.

6 Bourne, A., Cunningham, S., Spottiswoode, C. and Ridley, A. (2020) Hot droughts compromise interannual survival across all group sizes in a cooperatively breeding bird. *Ecology Letters* 23: 1776–88.

7 Ridley, A. R., Wiley, E. M., Bourne, A. R., Cunningham, S. J. and Nelson-Flower, M. J. (2021) Understanding the potential impact of climate change on the behavior and demography of social species: the pied babbler (*Turdoides bicolor*) as a case study. *Advances in the Study of Behaviour* 53: 225–66.

Chapter 31

1 Lees, A. C. and Peres, C. A. (2006) Rapid avifaunal collapse along the Amazonian deforestation frontier. *Biological Conservation* 133: 198–211.

2 Gardner, T. A., Ferreira, J., Barlow, J., Lees, A. C., *et al.* (2013) A social and ecological assessment of tropical land uses at multiple scales: the Sustainable Amazon Network. *Philosophical Transactions of the Royal Society B: Biological Sciences* 368: 20120166.

3 Nepstad, D., McGrath, D., Stickler, C., Alencar, A., Azevedo, A., *et al.* (2014) Slowing Amazon deforestation through public policy and interventions in beef and soy supply chains. *Science* 344: 1118–23.

4 Lewis, S. L., Brando, P. M., Phillips, O. L., van der Heijden, G. M. and Nepstad, D. (2011) The 2010 Amazon drought. *Science* 331: 554.

5 Barlow, J., Lennox, G. D., Ferreira, J., Berenguer, E., Lees, A. C., *et al.* (2016) Anthropogenic disturbance in tropical forests can double biodiversity loss from deforestation. *Nature* 535: 144–7.

6 França, F. M., Benkwitt, C. E., Peralta, G., Robinson, J. P., Graham, N. A., Tylianakis, J. M., Berenguer, E., Lees, A. C., Ferreira, J., Louzada, J. and Barlow, J. (2020) Climatic and local stressor interactions threaten tropical forests and coral reefs. *Philosophical Transactions of the Royal Society B* 375: 20190116.

7 Stouffer, P. C., Jirinec, V., Rutt, C. L., Bierregaard Jr, R. O., Hernández-Palma, A., *et al.* (2021) Long-term change in the avifauna of undisturbed Amazonian rainforest: ground-foraging birds disappear and the baseline shifts. *Ecology Letters* 24: 186–95.

8 Barlow, J. and Lees, A. C. (2019, 23 August) Amazon fires explained: what are they, why are they so damaging, and how can we stop them? *The Conversation*. https://theconversation.com/amazon-fires-explained -what-are-they-why-are-they-so-damaging-and-how-can-we-stop -them-122340

9 Barlow, J., França, F., Gardner, T. A., Hicks, C. C., Lennox, G. D., *et al.* (2018) The future of hyperdiverse tropical ecosystems. *Nature* 559: 517–26.

10 https://ebird.org/

Afterword

1 Smith, S. R., Christie, I. and Willis, R. (2020) Social tipping intervention strategies for rapid decarbonisation need to consider how change happens. *PNAS* 20: 10629–10630.

Index

Also available from Pelagic

Wild Mull: A Natural History of the Island and its People, by Stephen Littlewood and Martin Jones

The Hen Harrier's Year, by Ian Carter and Dan Powell

101 Curious Tales of East African Birds, by Colin Beale

Urban Peregrines, by Ed Drewitt

The Wryneck, by Gerard Gorman

Rebirding: Restoring Britain's Wildlife, by Benedict MacDonald

A Field Guide to Harlequins and Other Common Ladybirds of Britain and Ireland, by Helen Boyce

A Natural History of Insects in 100 Limericks, by Richard A. Jones and Calvin Ure-Jones

Essex Rock: Geology Beneath the Landscape, by Ian Mercer and Ros Mercer

Bat Calls of Britain and Europe, edited by Jon Russ

Pollinators and Pollination, by Jeff Ollerton

Rhythms of Nature: Wildlife and Wild Places Between the Moors, by Ian Carter

A Miscellany of Bats, by M. Brock Fenton and Jens Rydell

Ancient Woods, Trees and Forests: Ecology, History and Management, edited by Alper H. Colak, Simay Kırca and Ian D. Rotherham

www.pelagicpublishing.com